RESEARCH METHODOLOGY IN AGRICULTURAL EXTENSION

The Authors

Dr. C. Karthikeyan, Professor and Head, Dept. of Agricultural Extension and Rural Sociology, TNAU, Coimbatore has more than 27 years of professional experience. Specialised in Cyber extension and outreach. Principal Investigator for World Bank Funded M-Velanmai Project. Proposed the concept of ICT based Extension approaches namely, 'e-Velanmai and M-Velanmai through World Bank support. Won Eleven International recognitions/awards and 20 National/ State level awards for his contributions. Guided 18 M.Sc. (Ag. Extension), 14 M.B.A. and 8 Ph.D. (Ag. Extension) students. He was recognized as a Fellow by the Indian Society of Extension Education, New Delhi. Published 231 books, 40 book chapters, 34 technical books & 249 Scientific articles. Completed 33 externally funded projects with the budget of Rs. 28.34 Crores. Visited 8 Countries namely USA, Australia, Italy, Sweden, Egypt, Bangladesh, China and Nepal.

Dr. C. Cinthia Fernandaz, Completed her PG and Ph.D in Agricultural Extension from Tamil Nadu Agricultural University, Coimbatore. Specialized in the Field of Forestry Extension at Forest College and Research Institute, Mettupalayam, she has 10 years of experience in Teaching, Research and Extension. She is a recipient of President of India Award for Best Thesis for Poverty Alleviation by TNAU for Ph.D research. She contributes as PI and Co-PI in several externally funded projects in the Field of Forestry and Agroforestry. She is also involved in conducting PRA towards documentation of base line analysis. She published 20 research papers, five books, manuals, technical reports, technical articles and 30 popular articles. She has conducted several awareness campaigns, exhibitions and capacity building programmes to various stake holders in the field of Forestry and Agroforestry. She is coordinating the Training activities of TNAU under Training Division.

Dr. P. Jaisridhar is a distinguished Assistant Professor at the Tamil Nadu Agricultural University (TNAU), Coimbatore, with a specialization in Agricultural Extension. His extensive career spans multiple roles including the current role as Assistant Professor in the ICAR - KVK, The Nilgiris and service at various research and educational institutes such as the Centre of Excellence in Millets, Tiruvannamalai, Horticultural College and Research Institute for Women, Trichy, ICAR-Krishi Vigyan Kendra, Vellore and Directorate of Extension Education, TNAU, Coimbatore. Dr. Jaisridhar has delivered a variety of undergraduate and postgraduate courses, organized educational tours, and guided PG students for postgraduate research. His scholarly contributions include six books, several chapters, and a plethora of research papers and e-learning resources. Recognized for his Extension services & research excellence, he has garnered multiple international and national awards. His dedication to agricultural innovation and extension is reflected in his involvement in numerous externally funded projects and his leadership in organizing training programs and workshops for farmers and students alike.

RESEARCH METHODOLOGY IN AGRICULTURAL EXTENSION

– Authors –

Dr. C. Karthikeyan

Professor and Head
Department of Agricultural Extension and Rural Sociology
Tamil Nadu Agricultural University
Coimbatore

Dr. C. Cinthia Fernandaz

Associate Professor (Agricultural Extension)
Directorate of Extension Education
Tamil Nadu Agricultural University
Coimbatore

Dr. P. Jaisridhar

Assistant Professor (Agricultural Extension)
Tamil Nadu Agricultural University
ICAR - KVK, Nilgiris

2026

Daya Publishing House®
A Division of
Astral International Pvt. Ltd.
New Delhi – 110 002

ISBN 978-93-5919-574-2

Published by : **Daya Publishing House®**
A Division of
Astral International Pvt. Ltd.
– ISO 9001:2015 Certified Company –
4736/23, Ground Floor, Ansari Road
Darya Ganj, New Delhi-110 002
Ph. 011-43549197, 23278134
e-mail: info@astralint.com
Website: www.astralint.com

Printed at : **Replika Press Pvt Ltd**

Preface

The book is designed to introduce students to the process of scientific research in Agricultural Extension. This book is based on the class lectures and experiences gained during the past years at the Department of Agricultural Extension and Rural Sociology. The target groups includes students, researchers and a ready hand book for Professors and Scientists. The target group for this book includes students, junior researchers, and professors teaching courses on research methods.

This book is not just about "research methods" (empirical data collection and analysis) but about the entire "research process" from start to end. Research method is only one phase in that research process, and possibly the easiest and most structured one. The chapters are devoted to develop the skills of Researcher" and "Scientific Research", which are essential skills for a junior researcher.

The entire book is designed compact and focuses on essential. Most courses include a fair complement of readings drawn from the respective discipline. This book is designed to complement those readings by summarizing all important concepts. .

The book is structured into 30 chapters for a 16- week semester containing the entire syllabus. These chapters includes research methods, Science, social science research, types of knowledge, research problem, sampling, Quantitative, Qualitative research, Research questions, literature review, behavioral research, variables, hypothesis, measurement, designs, strategies, research methods, data collection, item analysis, reliability, validity, data coding, editing, statistical analysis, interpretation of data, citation and citation manager. Nevertheless, there are chapters on statistical analysis for research.

It is planned to continually update this book based on emerging trends in scientific research. If there are any new or interesting content that you wish to see in future editions, please drop a note, and we will try our best to accommodate them. Comments, criticisms, or corrections to any of the existing content will also be gratefully appreciated.

Dr. C. Karthikeyan

Dr. C. Cinthia Fernandaz

Dr. P. Jaisridhar

Contents

Chapter 1

Science and Common Sense

Science

The systemic study of the structure and behaviour of the physical and natural world through observation, experimentation, and the testing of theories against the evidence obtained.

Science is a body of verified knowledge.

Science is a method/study/approach leads to systematic knowledge about how things operate.

It is a future predictions based on the past experiences.

The term science may be thought of as an approach to the gathering of knowledge rather than as a field of subject matter. Science, put simply, consists of two primary functions:(1) the development of theory and (2) the testing of substantive hypotheses that are deduced from theory.

Common Sense

The ability to think and behave in a reasonable way and to make good decisions.

Common sense is sound, practical judgement concerning everyday matters, or a basic ability to perceive, understand, and judge in a manner that is shared by nearly all people.

Difference Between Science and Common Sense

FEATURES	*SCIENCE*	*COMMON SENSE ON SENSE*
Conceptual scheme and theoretical functions	Scientist use them	Layperson may also use them but does so in loose fashion accepting fanciful explanation of natural and human phenomenon
Theories and hypotheses	Scientist test them systematically and empirically	Non scientist will also test them but in a selective fashion

Notion of control	Scientist try to systematically rule out variables that are possible causes of the effects	Lay people seldom control their explanation of observed phenomena systematically
Preoccupation with relationship	Scientist consciously and systematically pursues relationships	Lay persons preoccupation with relationship is loose unsystematically and uncontrolled
Different explanation of observed phenomena	Carefully rules out metaphysical explanation	Much more concerned with metaphysical explanation

Four Methods of Knowing

- Method of tenacity – Hold firmly on truth, frequent repetition of such truths seems to enhance its validity
- Method of authority – Holds on established beliefs – An idea has weight of tradition / public sanction
- Method of priori (Intuition) – Agree without any past experience/observation – Accept by intuition
- Method of science – Common for all – verifiable - self correlation

Inductive Approach: (Inductive Reasoning)

Inductive approach involves the search for pattern from observation and the development of explanation and theories for those patterns through series of hypotheses.

No theories or hypotheses would apply in inductive studies at the beginning of the research and the researcher is free in terms of altering the direction for the study after the research process had commenced.

Deductive Approach: (Deductive Reasoning)

Concerned with developing a hypotheses based on existing theory, and then designing a research strategy to test the hypotheses.

Research

The word research is originated from a French word " recerche" which means to search/ critical examination into a topic or problem .

According to Webster's dictionary research is careful or diligent search ; studio inquiry ; usually critical and exhaustive investigation ; aimed at the discovery and interpretation of facts ; revision of accepted theories or laws in the light of new facts or practical application of such new or revised theories or laws.

- Research commonly refers to a search for knowledge
- It is a scientific and systematic investigation for some pertinent information on a specific topic.

- Systematic, controlled, empirical and critical investigations of hypothetical propositions about presumed relations among natural phenomenon. (Kerlinger,1984)
- According to Best (1977) RESEARCH is considered to be the more formal, systematic, and intensive process of carrying on a scientific method of analysis. It is a more systematic activity directed toward discovery and the develop of an organized body of knowledge.
- It is a more systematic activity directed toward discovery and the development of an organized body of knowledge.

Importance of Research

- Inseparable part of human knowledge
- Important pre- requisite for dynamic social order
- All researches are essentially socially oriented
- Careerism for PG students
- Direct outcome of man's urge to understand the facts
- Research methodology is the source of livelihood for scientists / professionals
- For philosophers and thinkers- it means development of new styles and creative work
- For analysts and intellectuals – it means the generalization of new theories

Characteristics of Research

- Research require a strong determination to know the unknown
- Research is characterized by patient and unhurried activity
- Research is directed toward the solution of problem
- Research is based upon observable experience or empirical evidence
- Research requires expertise
- Research emphasis the development of generalization, principle, theories that will be helpful in predicting future
- Research involves gathering new data from primary or first hand sources or using existing (secondary) sources data for new purpose
- Research requires expertise
- Research strives to be objective and logical , applying every possible test to validate the procedures employed, the data collected , and the conclusion reached
- Research is carefully recorded and reported

Purpose of Research

- Explorative purpose: Exploratory research is the first research to be conducted around a problem that has not yet been clearly defined.

Exploration research therefore aims to gain a better understanding of the exact nature of the problem and not to provide a conclusive answer to the problem itself. This enables us to conduct more in- depth research later on.

- Descriptive purpose: Descriptive research expands knowledge of a research problem or phenomenon by describing it according to its characteristics and population. Descriptive research focuses on the 'how' and 'what', but not on the 'why'.
- Explanatory purpose: Explanatory research, also referred to as casual research, is conducted to determine how variables interact, i.e. to identify cause-and-effect relationships. Explanatory research deals with the 'why' of research questions and is therefore often based on experiments.

Characteristics of a Good Researcher

- Curiosity: Good researchers constantly strive to learn more, not just about their own field, but about other fields as well.
- Analytical ability and foresight: Good researchers evaluate data from every angle and search for patterns
- Determination: A good researcher accepts the disappointments, learn from the failure, re-evaluate their experiment, and keep moving forward.
- Collaboration: Research is becoming increasingly multidisciplinary. Being able to coordinate, and get along with team members is a major factor that can contribute to one's success as a researcher.
- Communication: Communication skills are an essential asset for every researcher. Not only do they have to communicate with their team members, but they have to communicate with co-authors, journals, publishers, and funders

Scientific Method

It is a process with the help of which scientists try to investigate, verify, or construct an accurate and reliable version of any natural phenomenon.

Done by creating an objective framework for the purpose of scientific enquiry.

Characteristics of Scientific Method

- Objectivity: The procedure and the findings are not changed by the researcher feeling , values or beliefs after the research begun . Objectives are those of a detached , impersonal ,unbiased or un prejudicial nature
- Generality: Generality means the ability to generalize from particular phenomenon under study. Conclusion or generalizations drawn should be of universal application
- Verifiability: Verifiability is a characteristic which indicates that the observed phenomena or facts can be verified by subsequent studies

- Predictability: This means that the results of the study can predict the future events with sufficient accuracy.

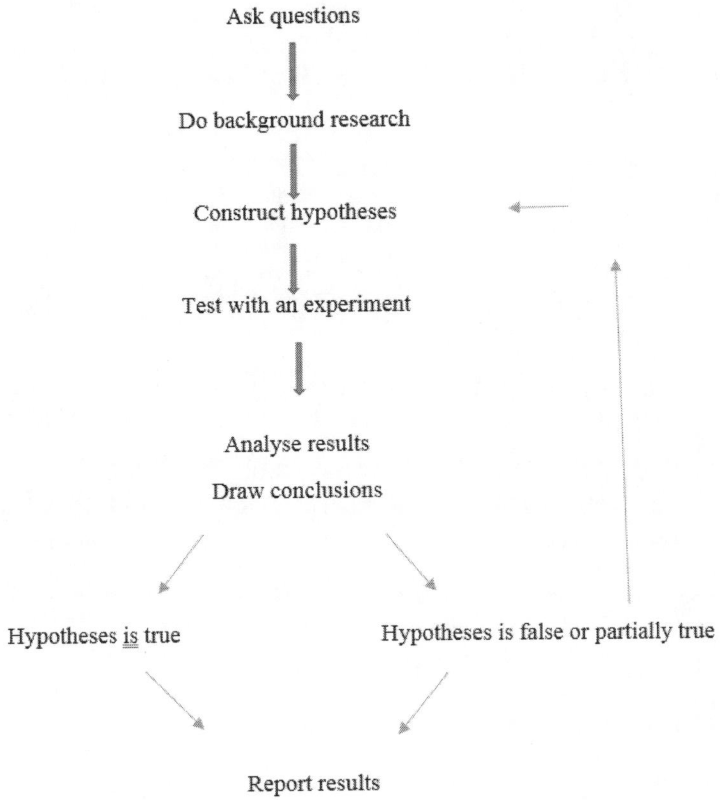

Stages in Scientific Approach

- Formulation of problems and hypothesis
- Formulation and application of methods of data collection
- Classification and interpretation of results
- Generalization and formulation of laws

References

Research methodology- Technology and Trends Vol. 1- Singh (Y.K) and Bajpai (R.B)

Agricultural Extension- Innovation and Dimensions- Dileep Kumar and Khajan Singh

Extension Education- Core Contents and Emerging Areas

https://www.discoverphds.com

https://www.researchmethod.net

Chapter 2

Types of Social Research

Science is the concerned with prediction of events in the physical and biological universe, scientific thoughts is a body of facts and methods and models

(Marshal Walker)

Science is distinguished from common sense knowledge by the rigor with which it subordinates all other consideration to the pursuit of the ideal of certainty, exactness, universality and system

(Wilson Gee)

Steps in Conducting Research

The following general steps are involved in conducting a research :

- Formulating the research problem
- Survey and review of literature
- Developing hypothesis and clarifying the concepts
- Deciding on the research design
- Collecting data with the help of tools and techniques already available and /or specially designed for the purpose
- Analysis and interpretation of data
- Testing the hypothesis
- Conclusions , implication , recommendation
- Preparation and submission of report

A research finding is considered as credible and adds to our knowledge only when a proper and scientific research methodology is adopted.

Six Characteristics of Science

Conptt:

Consistency	Several scientists can repeat an experiment and the results are the same.
Observability	The results or subject of an experiment can be observed by using the five senses or extensions of them.

Natural	A natural cause must be used to explain why or how the naturally. occurring result or experimental subject happens. Supernatural can't be used.
Predictability	Predictions can be made about the subject and can be tested.
Testability	Subject can be tested using a controlled experiment.
Tentativeness	Scientific theories are subject to change because of technology.

Behavioural Science

Behavioural science describes the study of human behaviour through the use of systematic experimentation and observation. Behavioural scientists study when and why individuals engage in specific behaviours by experimentally examining the impact of factors such as conscious thoughts, motivation, social influences, contextual effects, and habits. Several disciplines fall under the broad label of behavioural science, including:

- Anthropology
- Behavioural economics
- Cognitive psychology
- Consumer behaviour
- Social psychology
- Sociology

In order to understand the full complexity of human behaviour, some behavioural scientists synthesize theories, concepts, and methodologies across some of these disciplines. For example, the field of behavioural economics emerges from bringing insights from psychology to bear on economic behaviour, thereby predicting and explaining behaviour that is not anticipated by standard economic theories.

Behavioural science research is diverse and expansive. Behavioural scientists study why humans sometimes behave in a way that may not maximize their own well-being, such as making choices in the present that do not maximize their happiness in the future; examine how seemingly arbitrary contextual factors influence our decisions, beliefs, and attitudes; test how different incentives affect people's motivation and behaviour; analyse how people judge others' traits and characteristics based on features of their face or voice; investigate how consumers can be encouraged to make, avoid, or change spending decisions; and design policy interventions that can help people make choices that they would personally recognize as optimal in the long run.

Social Science Research

Meaning and Scope

Sciences are broadly divided into natural (physical) sciences and social sciences. Social sciences include various disciplines dealing with human life, human behaviour and institutions. e.g. Anthropology, History, Economics, Education, Commerce, Demography etc,. Social sciences are not exact science like physical sciences. It deals with human beings. Human nature and mans environment are so complex, that it is more difficult to comprehend and predict human behaviour than the physical phenomena. It is difficult see the underlying uniformity in the diversity of complex human behaviour.

Social Science Research

Social science research is a systematic method of exploring, analysing and conceptualizing human life in order to extend, correct or verify knowledge of human behaviour and social life. Social research seeks to find explanations to unexplained phenomena, to clarify the doubtful and correct the misconceived fact of social life. It involves the application of scientific method for understanding and analysing of social life in order to correct and verify the existing knowledge as a system. The main idea behind social research is to discover new inter relations, new knowledge, new facts and also to verify old ones. Human behaviour may be involved by certain values and laws. The main purpose of social research is to discover those laws which can be proper guidelines for studying human contact and behaviour. According to P.V. Young, we may define social research as "the systematic method of discovering new facts and verifying old facts. Their sequences inter relationship, causal explanations and the natural laws govern them". From the above definition we can identify the following.

Functions of Social Science Research

The important functions of social science research are discussed below;

a) Discovery of facts and their interpretation: Social research provides answer to questions of what, when, how and why of man, social life and institutions. Discover of facts and their inter relationship help us to discard distortions and contribute to our understanding of social reality.

b) Diagnosis of problems and their analysis: Our society has innumerable problems such as poverty, unemployment, economic inequality, social tension etc,. The nature and dimensions of such problems have to be diagnosed and analyzed. An analysis of problems leads to an identification of appropriate remedial actions.

c) Systematization of knowledge: The facts discovered through research are systematized and the body of knowledge is developed. It contributes to the growth of theory building.

d) Control over social phenomena: Research in social science provides first hand information about the nature of social institutions. This knowledge helps us to control over the social phenomena.

e) Prediction: Social research aims at finding an order among social fact and their casual relations. This affords a sound basis for prediction in several cases

f) Development planning: Systematic research can give us the required data base for planning and designing developmental schemes and programmes.

g) Social welfare: Social research can identify the causes of social evils and problems. It can thus help in taking appropriate remedial actions. It also provides guideline for social welfare.

Scope of Social Science Research

The fields of social science research unlimited and the materials of research are endless.Every group of social phenomena, every phase of human life and every stages of past and present development are materials for the social scientist. The area of research in various social sciences provides vast scope for research in social sciences.

Types of Research

Basic or Fundamental Research:

To gather knowledge for the sake of knowledge. It is concerned with solving any practice problem, but with discovering underlying and if possible, universal laws and theories .

- **Applied Research or practical research:** Gathering of knowledge that could aid in the betterment of human destiny .
- **Action Research:** It is a form of social research which aims at better insight into the problem by learning from the experience gained in attempt to solve the problems .

Types of Social Research

There are four main types of Social Research: Qualitative and Quantitative Research, Primary and Secondary Research.

Qualitative Research: Qualitative Research is defined as a method to collect data via open-ended and conversational discussions, There are five main qualitative research methods- ethnographic research, focus groups, one-on-one online interview, content analysis and case study research. Usually, participants are not taken out of their ecosystem for qualitative data collection to gather information in real-time which helps in building trust. Researchers depend on multiple methods to gather qualitative data for complex issues.

Quantitative Research: Quantitative Research is an extremely informative source of data collection conducted via mediums such as surveys, polls, and questionnaires. The gathered data can be analyzed to conclude numerical or statistical results. There are four distinct quantitative research methods: survey research, correlational research, causal research and experimental research. This research is carried out on a sample that is representative of the target market usually using close-ended questions and data is presented in tables, charts, graphs etc.

For example, A survey can be conducted to understand Climate change awareness among the general population. Such a survey will give in-depth information about people's perception about climate change and also the behaviours that impact positive behaviour. Such a questionnaire will enable the researcher to understand what needs to be done to create more awareness among the public.

Primary Research: Primary Research is conducted by the researchers themselves. There are a list of questions that a researcher intends to ask which need to be customized according to the target market. These questions are sent to the respondents via surveys, polls or questionnaires so that analysing them becomes convenient for the researcher. Since data is collected first-hand, it's highly accurate according to the requirement of research.

For example: There are tens of thousands of deaths and injuries related to gun violence in the United States. We keep hearing about people carrying weapons attacking general public in the news. There is quite a debate in the American public as to understand if possession of guns is the cause to this. Institutions related to public health or governmental organizations are carrying out studies to find the cause. A lot of policies are also influenced by the opinion of the general population and gun control policies are no different. Hence a gun control questionnaire can be carried out to gather data to understand what people think about gun violence, gun control, factors and effects of possession of firearms. Such a survey can help these institutions to make valid reforms on the basis of the data gathered.

Secondary Research: Secondary Research is a method where information has already been collected by research organizations or marketers. Newspapers, online communities, reports, audio-visual evidence etc. fall under the category of secondary data. After identifying the topic of research and research sources, a researcher can collect existing information available from the noted sources. They can then combine all the information to compare and analyse it to derive conclusions.

Importance of Social Research

Source of Knowledge. Social research is an important source of knowledge. It opens new ways of knowledge and wisdom.When something unknown brings to a researcher, there is a sort of inner pleasure and happiness. It gives knowledge to the researcher about the unknown facts. It paves out the way of ignorance and gives new direction in a social life.

Research in Informative. It provides updated, proved, authentic and very useful information to the end users. Researchers and common people can take advantages from research. It is not only informative. Its findings are scientific based therefore the results and information are reliable.

Social Cohesion and Unity. Social unity in a problematic area b/w two groups or castes is only possible due to social research. Social research manifests the causes of these events and gives a remedial solution to it. So, social researcher is the main factor for bringing social coherence and unity in society.

Social Planning. If someone wants to develop and bring social growth and development in a society, social planning will be developed which is the outcome of social research. In this situation achieve the growth of society on right line is highly necessary. Social growth can possible only when problems are solved.

Solution of Social Problems. Social research also helps in the solution of social problems it is the research through which we can find the causative factors of an existing social problem and guide us about his solution.

Structural Changes. Social research is responsible for bringing structural changes in a social life. We bring social and cultural change in a social situation which is the outcome of social research.

Social Welfare. Social evils can be removed from society due to research. The process of social welfare is possible due to the procedure of social research through which social evils can be removed.

Social Prediction. Social research helps in making laws and the relation between social laws and their factors. It helps to know the existence of causative factors and their magnitude and this facilitates prediction of results.

Social Control. For smooth running of society social control is necessary. For this purpose we apply social research to a society for bringing social order and control.

Social Growth. The growth and development in a society is possible for the improvement of living standard and it is the main function of social research.

New Techniques. Social Research introduces new techniques and improves the old ones. It also modifies the existing tools of research. It gives us the methods of knowing new ways for using our research procedure.

References

Prinicples of Research in Social Psychology -William D.Crano,Marilynn B.Brewer

Research Methodology Methods and Techniques -C R Kothari

https://www.questionpro.com/blog/social-research/

Social Science Research: Principles, Methods, and Practices - Anol Bhattacherjee

https://studylecturenotes.com/importance-of-social-research/

Advances and Challenges in Agriculture extension and Rural Development – T.Rathakrishnan, M.Israel Thomas ,Nirmala

Extension Education Core concepts and Emerging Areas – Debabrata Das Gupta.

Agricultural Extension , Innovations and Dimensions -Dileep Kumar , Khajan Singh .

Chapter 3

Types of Knowledge

Types of Knowledge Generated through Research

1. Historical knowledge
2. Axiological knowledge
3. Theoretical knowledge
4. Conceptual knowledge

1. Historical knowledge

- It studies the meaning of past events in an attempt to interpret the facts and explain the cause of events, and their effect in the present events.
- Historical knowledge refers to knowledge of historical events and it is factual, based on examination of historical documents.
- The purpose of historical knowledge to gather and produce knowledge of the recorded past.
- Historical knowledge refers to knowledge about past events, people and societies.
- Historical sources gives us an indepth understanding of things that have happened in the past from the point of view of the people who lived through them.

Objectives

- Explain how historians use the artifacts to find out what happened in the past.
- To understand that cultural background of the historian which may affect the interpretations.
- To understand how knowledge about history can shape personal and national identities.

Example:

- Reduces the biases towards a culture through understanding.
- Respecting cultural beliefs to establish better relationships with the society.

2. Axiological knowledge

- The word 'axiology' is derived from two Greek words 'axios' (worth or value) and 'logos' (logic or theory), meaning the theory of value. It is the branch of philosophy dealing with values, ethics, aesthetics, or religion .It is the study of value and value judgement .The science of Axiology enables us to identify our internal valuing systems and their influence on our perceptions, decisions and actions.
- Axiology is important because it can offer a more nuanced examination of lives
- It calls into question why we believe something is important or valuable and how we act or do not act based on that information ,which can lead to a more thoughtful existence.
- It is a study that can be applied or theorized in many different spaces .
- In this study ,one consider how and why one place judgement on something.

Examples;

- What values should one live by?
- Is morality defined by our actions, or by What is in our hearts?

3. Theoretical knowledge

- This helps us to understand why one technique works where another fails. It shows you the whole forest, builds the context, and helps you set strategy. Where self-education is concerned theory prepares you to set direction for your future education. Theory teaches us through the experience of others. This knowledge can often lead to a deeper understanding of a concept through seeing it in context of a greater whole and understanding the why behind .Theoretical knowledge means learning anything without adopting practical approach.

Two kinds of theoretical knowledge are sought in a design study .

- The first is a descriptive explanation of the processes operation in a domain,
- The second is technological or design knowledge about how to create and implement the technological tools that control the treatment effects.

4. Conceptual knowledge

- Conceptual knowledge refers to the knowledge of or understanding of concepts, principles, theories, models, classification etc.,
- Conceptual knowledge allow to us to collect facts and idea and group them into clusters that represents connectivity.
- Conceptual learning is the latest educational approach that focuses more on understanding the concepts and learning how to organize and distribute data.
- "Explicit or implicit understanding of the principles that govern the domain

and of interrelations between pieces of knowledge in a domain," – Rittle-Johnson & Wagner (1999).

- "Ideas, relationships, connections, or having a 'sense' of something." – Barr, Doyle et. El. (2003).
- "Learning that involves understanding and interpreting concepts and relations between concepts." – Arslan (2010).
- "To know why something happens in a particular way." – Hiebert and Lefevre (1986).

Examples:

- Classification and caterogies -Time period , music style.
- Principles and generation -Pythogerous theorem , inflation , gravity.
- Theories ,model and structure -Maslow's hierarchy of theory, evolution theory ,carbon cycle.

Prior Research Studies

- Prior studies allow the researcher to see and discuss the recommendation made by previous researchers.
- Previous research plays an important function in offering a vast number of sources and references for the researchers research.
- Prior knowledge is the information that one has before they acquire new ideas or concepts .

This information could have been gathered over time from their personal , social and culturals influences.

- It's basically understanding the work and research which has been followed by various scientists and researchers in the past.
- This gives the reader an overview of past research in the given field. This should positively end the research gaps such that new scope can be derived out of it. It helps to formulate and refine your research question. It avoids repetition.
- It helps you formulate a better argument.
- Prior knowledge helps in designing and modifying pedagogical practices.
- Prior knowledge makes learning meaningful.
- Prior knowledge facilitates integration of new knowledge.

Literature Review

A literature review is a piece of academic writing demonstrating knowledge and A literature review also includes a critical evaluation of the material ,this is why called as literature review

A literature review also includes understanding of the academic literature on specific topic placed in context.

- Its importance stems from the questions that you are trying to answer as part of your thesis.
- Usually, in a thesis document we will organize
  - what we are trying to do
  - why it is so important.
  - if it is so important, why has it not been done already?

Objective

- What are the other related items that are important also but we are not going to do this in this thesis (scope) It provides context to the content of the dissertation?
- It establishes credibility. Knowing that the author has adequate understanding of what others have done, what the new things can be done given everything is known, builds some confidence in the original work that follows.
- It collects a lot of relevant information in one place. The author's work can then directly build on it. If this survey of the literature is absent, the non-expert reader may not be able to make sense of dissertation.

Academic Debate

Debate: The process of inquiry and advocacy , the seeking of a reasoned judgement on a proposition .

Academic debate is conducted under the direction of an educational institution to provide educational opportunities for students.

- "Great leaders are almost always great simplifiers, who can cut through argument, debate and doubt, to offer a solution everybody can understand." – Colin Powell.
- It describes and discusses how ideas have changed over time.
- In chronological order, it examines the landmark studies that have moved the debate forward.
- It ends up where you start your own research question; it frames your contribution to the field.

Objectives:

- To train in argumentation.
- To empower personal expression.
- To become a truly effective communicator by becoming an effective listener.
- To understand the role and value of feedback.
- To appreciate the vital learning that can arise from feedback.

Reference

Advances and challenges in agricultural extension and rural development -T. Rathal, M. Israel Thomas and L. Nirmala.

Research Methodology (3rd edition) Methods and Techniques-C.R.Gaurav.

Btrezing ,B.H.,Kulhary.R.W(1979), Note taking and depth of processing, conemporary educational psychology , 4, no 2.

www.tltt.strath.ac.uk/REAP/public/Resources/DN .

Short answer

1. List out types of knowledge generated through research.
2. Define conceptual knowledge.
3. Define historical knowledge.
4. Mention some objectives of historical knowledge.
5. What is academic debate?
6. Explain Prior research.
7. What is axiological knowledge?
8. Explain literature review.
9. Explain theoretical knowledge.
10. Write down objectives of literature review.
11. List out objectives of academic debate.
12. Differentiate Historical and Axiological knowledge.

Fill in the blanks:

1. **Historical** knowledge studies the event of past.
2. The axiology derived from **Greek** word.
3. **Theoretical knowledge** helps to understand why one techniques works why another fails.
4. **Prior research** basically understanding the work and research which has been followed by various scientists and researchers in the past.
5. **Theoretical knowledge** can often lead to a deeper understanding of a concept through seeing it in context of a greater whole and understanding the why behind it.
6. **Academic debate** describes and discusses how ideas have changed over time.
7. **Axiological knowledge** is the study of value and value judgement .
8. **Literature review** importance stems from the questions that you are trying to answer as part of your thesis.
9. **Academic debate** helps you formulate a better argument.

10. **Axiology** enables us to identify our internal valuing systems and their influence on our perceptions, decisions and actions.
11. **Academic debate** is conducted under the direction of an educational institution to provide educational opportunities for students.

Chapter 4

Role of Behavioural Research in Extensions

Definitions

Behavioural research is a way to examine and understand individual and social behaviour through measurement and interpretation.

According to **P.V Yong** *"Social research is a systematic method of discovering the new facts or verifying the old fact, their sequences, inter-relationships, casual explanations and the natural laws which govern them".*

According to **C.A Moser**; *"Social research is the systematized investigation to gain new knowledge about social phenomena and surveys".*

Role of Behavioural Research

- It has to change the attitude of the farmer towards adoption of improved techniques of agriculture.
- It should adopt new methods of study of the farmer's problem which may lead to the formulation of social laws for presenting a constructive programme of social laws.
- It should conduct studies which aim at the collection and analysis of data on the various aspect of rural economy, so as to provide basic information for better and full utilisation of resources of the nation.

Behavioural Scientists: Behavioural scientists are professionals who research human behaviour and decision making as it relates to the society. They use their knowledge and expertise to conduct informed research, focusing on studying how people's actions affect their relationships with other.

Basic steps of research process:

There are 8 basic steps to conduct a research. They are;

1. Identifying the problem
2. Reviewing the literature
3. Setting research question and objective
4. Choosing the study design
5. Deciding on the Sample design

6. Collecting data
7. Processing and analysing the data and,
8. Writing the report

Identifying the problem: A well-identified problem will lead the researcher to accomplish all-important phases of the research process, starting from setting objectives to the selection of the research method. Researcher need to identify both; Non-research Problem, and Research Problem.

Reviewing of Literature It is an integral part of the research process. It enables the researcher to formulate his problem in terms of the specific aspects of the general area of his interest that has not been so far researched. Through a proper review of the literature, the researcher may develop the coherence between the results of his study and those of the others.

Setting research questions, objectives and hypotheses:

An **objective** will precisely say what should be researched, to delineate the type of information that should be collected, and provide a framework for the scope of the study. The best expression of a research objective is a well-formulated, testable research hypothesis.

A **hypothesis** is an unproven statement or proposition that can be refused or supported by empirical data. Hypothetical statements assert a possible answer to a research question. The objectives, questions and hypothesis is finalized based on the research topic and analysing the available literatures also.

Choosing the study design: It is a master plan specifying the methods and procedures for collecting, processing, and analysing the collected data. There are four basic research designs that a researcher can use to conduct his or her study; **Survey, Experiment, Secondary data study** and **observational study**

Deciding on the sample design:

- A **population** is the total collection of elements about which we wish to make some inference or generalization.
- A **sample** is a part of the population, carefully selected to represent that population.
- **Sample design** refers to the methods to be followed in selecting a sample from the population and the estimating technique, via formula for computing the sample statistics.

Simple random sampling, stratified random sampling, cluster sampling and **systematic sampling** are examples of different sample designs used.

Collecting data: The gathering of data may range from simple observation to a large-scale survey in any defined population. There are many ways to collect data, depending on the objective, design and availability of time and personnel.

E.g.:

- Structured interview.
- Self-administered questionnaires.
- Telephone interviewing
- Use of secondary sources, such as the census, vital registration records, official documents, previous surveys, etc.
- Qualitative data are collected mainly through **in-depth interviews, focus group discussions,** and observational studies.

Processing and Analysing Data: Data analysis usually involves reducing accumulated data to a manageable size, developing summaries, searching for patterns, and applying statistical techniques for understanding and interpreting the findings in the light of the research questions. The techniques may range from simple graphical technique to very complex multivariate analysis depending on the objectives of the study, research design employed, and the nature of data collected.

Writing the report - Developing Research Proposal, Writing Report, Disseminating and Utilizing Results: The end goal of a scientific study is to interpret the results and draw conclusions. The entire task of a research study is accumulated in a document called a proposal. It systematically outlines the particular research methodology and details the procedures that will be utilized at each stage of the research process. A report is an excellent means that helps to establish the researcher's credibility. At a bare minimum, a research report should contain sections on:

- An executive summary
- Background of the problem
- Literature review
- Methodology
- Findings
- Discussion
- Conclusions &Recommendations

What are research skills?

- Research skills are our ability to find answer to a question or a solution to a problem.
- It include the ability to gather information about your topic, review that information and analyse and interpret the information in a manner that brings us to a solution.
- Research skills are essential for employee to have in the most positions and industries.
- Research skill includes; searching for information, attention to detail, taking the notes time management, problem solving and communication skills.

Searching for information: All research involves the search for credible information that you can analyse and use to arrive at an answer or solution. In the workplace, the information you search for will likely help you complete projects and add value to the company. For example, you may look for information on how to resolve a company issue or identify your customers' needs

Attention to detail: Having the ability to pay close attention to details is often an essential skill in the workplace. Your job responsibilities will likely involve following specific policies and procedures outlined by the company, which means you need to know the details of those policies and procedures. Your job duties may also require you to pay close attention to the information customers give you so you can make a sale or resolve a customer complaint

Taking notes: Note-taking skills involve the ability to gather the most important information being presented to you and summarize it in writing so you can reference it later. Knowing how to take good notes will help you in several ways throughout your career. You will need to take effective notes during training and meetings. You may also need to take notes during interviews, especially if you are the interviewer.

Time management: Time management skills refer to the ability to break down projects into manageable parts, make a deadline-oriented plan and complete each section of your plan by your scheduled deadline. Time management skills involve setting goals, planning, organizing, delegating and prioritizing tasks. Displaying strong time management skills in the workplace can help you complete tasks on deadline, accomplish your goals and advance your career.

Problem-solving: Problem-solving skills refer to the ability to handle difficult situations and overcome complex business challenges. Problem-solving skills involve the abil9ity to break a problem down into its parts, think critically about each element, analyse the information you find and use that information to form an effective solution. Having strong problem-solving skills will help make you a valuable asset to your company and can help you advance your career.

Communicating results: Communication skills refer to the ability to understand the information you receive and present information in a manner others can understand. Communication skills involve active listening, observing and speaking. Research skills require good communication skills to present the results of your research. Strong communication skills are often one of the most important skills employers look for in candidates

How to develop research skills

Everyone is a natural researcher. Whether you're looking for the best price on a product or researching a topic for a school project, you're using research skills. Follow these steps to expand upon those skills:

1. Make an outline.
2. Know your sources

3. Learn to use advanced search techniques
4. Practice

1. Make an outline

The first thing you should do when starting any research project is to make an outline to help guide your research. Your outline should include a plan for the questions you need to research and the information you need to arrive at a decision. It should also include a plan for how you will divide your research project into manageable parts and a schedule detailing when each piece of your project should be complete.

2. Know your sources

The internet has made locating information more accessible, but not every source is credible. It is important to know how to find reliable sources and analyse information to determine whether or not it is credible. To ensure you are gathering accurate information, try to verify information from one source by using another. Locate the original information source to verify its reliability.

3. Learn to use advanced search techniques

When you are using the internet for research, most search engines allow you to use advanced search preferences to customize your search results. These search techniques will help you find the information you are looking for from credible sources more efficiently

4. Practice

One of the best ways to develop any new or existing skill is to practice. You can practice developing research skills by creating small research projects for yourself to work on that can help you with a current life event.

For example, if you are planning a vacation, you can practice researching different activities and prices of hotels and transportation. If you are looking for a job, you can practice researching information about different positions and companies.

References

Debabrata Das.2019.Extension Education: Core, Contents and Emerging Areas

T. Rathakrishnan, L. Nirmala and M. Israel Thomas. 2009. Advances and Challenges in Agricultural Extension and Rural Development.

Dileep Kumar, Khajan Singh.2012. Agricultural Extension: Innovations and Dimensions.

Dipak De, Basavaprabhu Jirli.2013. A Handbook of Extension Education.

C.R.Kothari, Gaurav Garg.2019. Research Methodology: Methods and Techniques

SHORT ANSWERS

1. Definition on behavioural research.
2. What is the role of behavioural research?
3. What are the basic steps of research process?
4. Write notes on Review the literature.
5. Define Hypothesis
6. Write brief on identifying the problem?
7. What are research skills?
8. Define Time management.
9. How to develop research skills?
10. Define Problem – Skills.

FILL IN THE BLANKS

1. Behavioural scientists are professionals who research human behaviour and decision making as it relates to the society.
2. Review of the literature, the researcher may develop the coherence between the results of his study and those of the others.
3. Research objective is a well-formulated, testable research hypothesis.
4. A Hypothesis is an unproven statement or proposition that can be refused or supported by empirical data.
5. A Population is the total collection of elements about which we wish to make some inference or generalization.
6. A Sample is a part of the population, carefully selected to represent that population.
7. A Report is an excellent means that helps to establish the researcher's credibility.
8. Research skills are our ability to find answer to a question or a solution to a problem.
9. Problem-solving skills refer to the ability to handle difficult situations and overcome complex business challenges
10. One of the best ways to develop any new or existing skill is to practice.

Chapter 5

Behavioural Research Process

Behavioural research is a way to examine and understand individual and social behavior through measurement and interpretation. According to P.V Yong "Social research is a systematic method of discovering the new facts or verifying the old fact, their sequences, inter-relationships, casual explanations and the natural laws which govern them"and according to C.A Moser; "Social research is the systematized investigation to gain new knowledge about social phenomena and surveys"

Role of Social Science Research

- ❖ It has to change the attitude of the farmer towards adoption of improved techniques of agriculture.
- ❖ It should adopt new methods of study of the farmer's problem which may lead to the formulation of social laws for presenting a constructive programme of social laws
- ❖ It should conduct studies which aim at the collection and analysis of data on the various aspect of rural economy, so as to provide basic information for better and fuller utilisation of resources of the nation.

Behavioral scientists are professionals who research human behavior and decision making as it relates to the society. They use their knowledge and expertise to conduct informed research, focusing on studying how people's actions affect their relationships with other

Basic Steps of Research Process

There are **8 basic step**s to conduct a research. They are;

1. Identifying the problem
2. Reviewing the literature
3. Setting research question and objective
4. Choosing the study design
5. Deciding on the Sample design
6. Collecting data
7. Processing and analysing the data and,
8. Writing the report

Identifying the problem:

A well-identified problem will lead the researcher to accomplish all-important phases of the research process, starting from setting objectives to the selection of the research method. Researcher need to identify both; Non-research Problem, and Research Problem.

Reviewing of Literature

It is an integral part of the research process. It enables the researcher to formulate his problem in terms of the specific aspects of the general area of his interest that has not been so far researched. Through a proper review of the literature, the researcher may develop the coherence between the results of his study and those of the others.

Setting research questions ,objectives, and hypotheses,

An **objective** will precisely say what should be researched, to delineate the type of information that should be collected, and provide a framework for the scope of the study. The best expression of a research objective is a well-formulated, testable research hypothesis.

A **hypothesis** is an unproven statement or proposition that can be refuted or supported by empirical data. Hypothetical statements assert a possible answer to a research question. The objectives, questions and hypothesis is finalized based on the research topic and analyzing the available literatures also.

Choosing the study design:

It is a master plan specifying the methods and procedures for collecting, processing, and analyzing the collected data. There are four basic research designs that a researcher can use to conduct his or her study; **Survey, Experiment**, **Secondary data study** and **observational study**

Deciding on the sample design:

A **population** is the total collection of elements about which we wish to make some inference or generalization. A **sample** is a part of the population, carefully selected to represent that population. **Sample design** refers to the methods to be followed in selecting a sample from the population and the estimating technique, vis-a-vis formula for computing the sample statistics. **simple random sampling**, **stratified random sampling**, **cluster sampling** and **systematic sampling** are examples of different sample designs used

Collecting data:

The gathering of data may range from simple observation to a large-scale survey in any defined population. There are many ways to collect data, depending on the objective, design and availability of time and personnel

Eg,

- Structured interview.
- Self-administered questionnaires.
- Telephone interviewing
- Use of secondary sources, such as the census, vital registration records, official documents, previous surveys, etc.
- Qualitative data are collected mainly through **in-depth interviews, focus group discussions, KII,** and observational studies.

Processing and Analysing Data:

Data analysis usually involves reducing accumulated data to a manageable size, developing summaries, searching for patterns, and applying statistical techniques for understanding and interpreting the findings in the light of the research questions. The techniques may range from simple graphical technique to very complex multivariate analysis depending on the objectives of the study, research design employed, and the nature of data collected.

Writing the report – Developing Research Proposal, Writing Report, Disseminating and Utilizing Results:

The end goal of a scientific study is to interpret the results and draw conclusions. The entire task of a research study is accumulated in a document called a proposal. It systematically outlines the particular research methodology and details the procedures that will be utilized at each stage of the research process. A report is an excellent means that helps to establish the researcher's credibility. At a bare minimum, a research report should contain sections on:

- An executive summary
- Background of the problem
- Literature review
- Methodology
- Findings
- Discussion
- Conclusions & Recommendations

What are research skills?

Research skills are our ability to find answer to a question or a solution to a problem.it include the ability to gather information about your topic, review that information and analyse and interpret the information in a manner that brings us to a solution. Research skills are essential for employee to have in the most positions and industries. Research skill includes; searching for information, attention to detail, taking the notes time management, problem solving and communication skills.

Searching for information:

All research involves the search for credible information that you can analyse and use to arrive at an answer or solution. In the workplace, the information you search for will likely help you complete projects and add value to the company. For example, you may look for information on how to resolve a company issue or identify your customers' needs.

Attention to detail:

Having the ability to pay close attention to details is often an essential skill in the workplace. Your job responsibilities will likely involve following specific policies and procedures outlined by the company, which means you need to know the details of those policies and procedures. Your job duties may also require you to pay close attention to the information customers give you so you can make a sale or resolve a customer complaint.

Taking notes:

Note-taking skills involve the ability to gather the most important information being presented to you and summarize it in writing so you can reference it later. Knowing how to take good notes will help you in several ways throughout your career. You will need to take effective notes during training and meetings. You may also need to take notes during interviews, especially if you are the interviewer.

Time management:

Time management skills refer to the ability to break down projects into manageable parts, make a deadline-oriented plan and complete each section of your plan by your scheduled deadline. Time management skills involve setting goals, planning, organizing, delegating and prioritizing tasks. Displaying strong time management skills in the workplace can help you complete tasks on deadline, accomplish your goals and advance your career.

Problem-solving:

Problem-solving skills refer to the ability to handle difficult situations and overcome complex business challenges. Problem-solving skills involve the abil9ity to break a problem down into its parts, think critically about each element, analyze the information you find and use that information to form an effective solution. Having strong problem-solving skills will help make you a valuable asset to your company and can help you advance your career.

Communicating results:

Communication skills refer to the ability to understand the information you receive and present information in a manner others can understand. Communication skills involve active listening, observing and speaking. Research skills require good communication skills to present the results of your research. Strong communication skills are often one of the most important skills employers look for in candidates.

How to develop research skills

Everyone is a natural researcher. Whether you're looking for the best price on a product or researching a topic for a school project, you're using research skills. Follow these steps to expand upon those skills:

1. Make an outline

The first thing you should do when starting any research project is to make an outline to help guide your research. Your outline should include a plan for the questions you need to research and the information you need to arrive at a decision. It should also include a plan for how you will divide your research project into manageable parts and a schedule detailing when each piece of your project should be complete.

2. Know your sources

The internet has made locating information more accessible, but not every source is credible. It is important to know how to find reliable sources and analyze information to determine whether or not it is credible. To ensure you are gathering accurate information, try to verify information from one source by using another. Locate the original information source to verify its reliability.

3. Learn to use advanced search techniques

When you are using the internet for research, most search engines allow you to use advanced search preferences to customize your search results. These search techniques will help you find the information you are looking for from credible sources more efficiently.

4. Practice

One of the best ways to develop any new or existing skill is to practice. You can practice developing research skills by creating small research projects for yourself to work on that can help you with a current life event. For example, if you are planning a vacation, you can practice researching different activities and prices of hotels and transportation. If you are looking for a job, you can practice researching information about different positions and companies.

References

Nakkiran, S., G. Ramesh (2009). Research Methods in Rural development. Deep & Deep Publications. Pp

Ganesan, R. (2011). Research Methodologies for Engineers. MJP Publishers. Pp 5-6

https://www.sciencedirect.com/topics/medicine-and-dentistry/behavioral-observation

Multiple Choice Questions

1. **What is the first step in behavioral research?**

   a) Review existing literature

   b) Formulate research questions

   c) Collect data

   d) Analyze data

2. **What does the literature review step in behavioral research involve?**

   a) Data collection

   b) Identifying research questions

   c) Reviewing existing research and literature

   d) Drawing conclusions

3. **Which step involves identifying the central questions to be addressed in your research?**

   a) Data analysis

   b) Defining variables

   c) Formulating research questions

   d) Hypothesis testing

4. **In the context of behavioral research, what are variables?**

   a) A specific type of data analysis

   b) Independent and dependent factors under investigation

   c) Research participants

   d) Literature sources

5. **Which step involves deciding on the methods and tools for data collection?**

   a) Data analysis

   b) Formulating hypotheses

   c) Choosing research methods

   d) Defining variables

6. **What are testable statements that predict outcomes in behavioral research called?**

   a) Variables

   b) Research questions

   c) Hypotheses

   d) Data

7. **When planning a study, what does the term "research design" refer to?**

   a) Identifying research questions

   b) Selecting data analysis methods

   c) The specifics of the research plan, including data collection procedures

   d) Choosing research methods

8. **What step involves the actual collection of data from research participants?**

   a) Formulating research questions

   b) Data analysis

   c) Collecting data

   d) Defining variables

9. **What is the process of interpreting and making sense of collected data?**

   a) Formulating hypotheses

   b) Literature review

   c) Data analysis

   d) Hypothesis testing

10. **What is the final step in behavioral research after data analysis?**

    a) Formulating research questions

    b) Reviewing existing literature

    c) Drawing conclusions and reporting findings

    d) Collecting data

Answers

1. b) Formulate research questions

2. c) Reviewing existing research and literature

3. c) Formulating research questions

4. b) Independent and dependent factors under investigation

5. c) Choosing research methods

6. c) Hypotheses

7. c) The specifics of the research plan, including data collection procedures

8. c) Collecting data

9. c) Data analysis

10. c) Drawing conclusions and reporting findings

Chapter 6

Research Problem and Research Topic

Definitions of Research Problem

A research problem, in general, refers to some difficulty which a researcher experiences in the context of either a theoretical or practical situation and wants to obtain a solution for the same. Thus, a research problem is one which requires a researcher to find out the best solution for the given problem, i.e., to find out by which course of action the objective can be attained optimally in the context of a given environment. There are several factors which may result in making the problem complicated. For instance, the environment may change affecting the efficiencies of the courses of action or the values of the outcomes; the number of alternative courses of action may be very large. Persons not involved in making the decision may be affected by it and react to it favorably of unfavorably, and similar other factors. All such elements (or at least the important ones) may be thought of in context of a research problem.

According to 'Kerlinger' the problem is an interrogative or probing statement where in, it will answer the relationship between two more variables.

According to 'Hurlock' "A problem then is an interrogative sentence or statement that asks what relation exists between two or more variables". e.g. Relationship between weather and fertilizer on yield of a crop.

Formulating the Research Problem

Quite often we all hear that a problem clearly stated is a problem half solved. This statement signify the need for defining a research problem. The problem to be investigated must be defined unambiguously for that will help to discriminate relevant data from the irrelevant ones. A proper definition of research problem will enable the researcher to be on the track whereas an ill-defined problem may create hurdles. Questions like: What data is to be collected? What characteristics of data are relevant and need to be studied? What relations are to be explored. What techniques are to be used for the purpose? and similar other questions crop up in the mind of the researcher who can well plan his strategy and find answers to all such questions only when the research problem has been well defined. Thus, defining a research problem properly is a prerequisite for any study and is a step of the highest importance. In fact, formulation of a problem is often more essential than its solution. It is only on the careful detailing of the research problem that we can work out the research design and can smoothly carry on all the consequential steps involved while doing research.

Techniques Involved in Defining Problem

- Statement of the problem in a general way.
- Understanding the nature of the problem.
- Surveying the available literature
- Developing the ideas through discussion
- Rephrashing the research problems into working proposition.

Sources of Research Problems

This section is of particular relevance if you have not yet selected a research topic and do not know where to start. If you have already selected your topic or question, go to the next section. Most research in the humanities revolves around four P's:

- People
- Problems
- Programs
- Phenomena

The emphasis on a particular 'P' may vary from study to study but generally, in practice, most research studies are based upon at least a combination of two Ps. You may select a group of individuals (a group or a community as such-'people'), either to examine the existence of certain issues or problems relating to their lives, to ascertain attitude of a group of people towards an issue ('problem"), to establish existence of a regularity (phenomenon") or to evaluate the effectiveness of an intervention ("program"). Your focus may be the study of an issue, an association or a phenomenon per se; for example, the relationship between unemployment and street crime, communities or organizations. The emphasis in these studies is on exploring, discovering or establishing associations or causation. Similarly, you can study different aspects of a program: its effectiveness, its structure, the need for it, consumers' satisfaction with it, and so on. In order to ascertain these you collect information from people. The 'people' provide you with the 'study population', whereas the other three Ps furnish the 'subject areas'. Your study population-individuals, groups and communities is the people from whom the information is collected. Your subject area is a 'problem', 'program' or 'phenomenon' about which the information is collected. A closer look at any academic discipline or occupational field will show that most research revolves around the four Ps. Every research study has two aspects:

1) the study population;

2) the subject area.

Characteristic of Good Research Problem

There are certain criteria to be followed in selection of a research problem. They are:

(i) The problem should be timely e.g.

(a) Scientist should think of inventing alternative methods to cooking fuel, since there is an increased demand for it and limited means of producing it. Solar energy can be tapped for this purpose which will be cheaper and feasible.

(b) Soon after good rain, light traps are arranged to eradicate outbreak of red hairy caterpillar pest.

(ii) The problem should relate to practical as well as theoretical situations.

e.g. (a) Green manuring in situ-gives better yields.

(b) High yielding varieties are grown for increasing yields.

(iii) The problem should permit generalisations to the maximum extent possible.

e.g. Mass media creates awareness about farm Technology in farmers.

(iv) The problem should help in developing new techniques, concepts and measurements.

e.g. The problem of how to apply seed and fertilizer at a time gave rise to new devices, seed-cum-fertilizer drill.

(v) The problem should allow to study the relationship of phenomena.

e.g. Study of Nature, Atmosphere, Weather.

(vi) Problem should express relationship between two or more variables.

E.g. The effect of rainfall and temperature on the yield of a crop.

(vii) Problem should be stated clearly unambiguous implying possibility of empirical testing.

e.g. Excess N' leads to excess vegetative growth.

Limitations of the Research Problem

The boundaries of the research problem which a student wants to investigate are limited by

(a) Time

(b) Resources

(c) Physical area to be covered

(d) Complexity as such, he cannot manage a broad problem which embodies many specific problems each requiring a separate in depth study.

In a new field of investigation, sometimes the problem may be of exploratory nature, the student feels hand- icapped in the absence of the availability of literature on the topic which he wants to study. Complexities of the problem also establish boundaries of the problem.

For instance, he would feel difficulty to undertake a problem related to determination of the influence of socio economic programmes on farm structure, employment pattern and level of income of the farmers. Even when a student is interested in investigating a problem relating to the impact of modern technology on cropping pattern and farm productivity, but he cannot extend his field of investigation to the whole of a given state because of constraints of time, energy and resources involved.

References

C R Kothari, Gaurav Garg. Research methodology methods and techniques. 3rd edition: New age international publishers.

Daivadeenam P.D, Somani L.L, Research methodology in Extension education. 2nd edition: Agrotech publisher.2004.

Ranjith kumar. Research methodology: A step by step guide for beginners. 2nd edition: Pearson Education Australia.2005

Questions

1. How do you define research problem? Give examples
2. Define the main issues which should receive the attention of the researcher in formulating the research problem. Give suitable examples
3. What is the necessity of formulating a research problem? Explain
4. Write a comprehensive note on the "Task of defining a research problem".
5. What are the sources of research problem?
6. Write a short note on characteristics if good research problem.
7. "The task of defining the research problem often follows a sequential pattern". Explain
8. Define research topic.
9. What are the techniques involved in defining the problem?
10. What the limitations of research problem?

Multiple Choice Questions

1) Research is

a) Searching again and again

b) Finding solution to any problem

c) Working in a scientific way to search for truth of any problem

d) None of the above

2) In order to pursue the research, which of the following is priorly required?

a) Developing a research design

b) Formulating a research question

c) Deciding about the data analysis procedure

d) Formulating a research hypothesis

3) A research problem is feasible only when

a) It has utility and relevance

b) It is new and adds something to knowledge

c) It is researchable

d) All of the above

4) What are those conditions where a research problem is not viable?

a) It is new and adds something to knowledge

b) It can be researched

c) It has utility and relevance

d) It contains dependent and independent variables

5) Out of the following, which characteristic does not pertain to the research problem?

a) Newness

b) Satisfaction of the mind of the researcher

c) Feasibility

d) Social significance

6) Formulation of a research problem depends upon:

a) What is the objective behind the researcher's choice?

b) What are the specific questions?

c) What is the conceptual model?

d) What negative factors to research upon?

e) Reasons for unlimiting the study.

f) Construction of hypothesis.

a) a, c, e and f

b) b, c, d and e

c) a, b, c and f

d) c, d, e and f

7) A research problem is feasible only when

a) It is researchable

b) It is new and adds something to knowledge

c) It has utility and relevance

d) All of these

8) Which of the following is not a criterion for the statement of a good research problem?

a) Expression of relationship between/among variables.

b) Clarity and unambiguousness.

c) Possibility of empirical testing.

d) Possibility of use of statistical analysis.

9) what is the purpose of doing research ?

a) To find problem

b) To find solution

c) Both a and b

d) None of the above

10) Which is not the source of research problem?

a) People

b) Problem

c) Program

d) Principle

Chapter 7

Quantitative and Qualitative Research

Research Problems

A research problem is a clear and concise statement that describes an issue or question to be addressed through research. It outlines the specific problem or gap in knowledge, emphasizes its significance, and provides the foundation for the research question and hypothesis.

- ❖ According to 'Kerlinger' the problems is an interrogative or probing statement Where in, it will answer the relationship between two more variables.
- ❖ According to 'Hurlock' "A problem then is an interrogative sentence or statement that asks what relation exists between two or more variables" Research problem is a statement about

✓ **Area of Concern** ✓ **Area to be improved** ✓ **A difficulty to be eliminated**

Personalized approach to selecting a research problem:

- ❖ Existing literatures
- ❖ Practitioners
- ❖ Brainstorming
- ❖ Popular issues
- ❖ Social concerns

Qualitative research

- ❖ Qualitative research methods are those generally subsumed under the heading ethnography.
- ❖ It includes **case studies, field studies, grounded theory, document studies, naturalistic inquiry, observational studies, interview studies and descriptive studies**.
- ❖ Qualitative research design in the social sciences stem from traditions in anthropology and sociology, where the philosophy emphasizes the phenomenological basis of a study, the elaborate description of the meaning of phenomena for the people or culture under examination.

- Qualitative is referred to as the **verstehen approach.**
- Focus of investigation is only one subject, one case, or one unit over an extended period of time.

Elements of qualitative research process:

Observation

↓

Patterns

↓

Hypothesis

↓

Theory

Quantitative research

- It falls under the category of **empirical studies** according to some, or **statistical studies, according to others.**
- These designs include more traditional ways in which psychology and behavioural science have carried out investigations.
- Quantitative modes have been the dominant methods of research in social science.
- Quantitative designs include experimental studies, quasi-experimental studies, control of variable and randomization.
- Data in qualitative studies are coded according to a priori operational and standardized definitions.

Elements of quantitative research process:

Theory

↓

Hypothesis

↓

Observation

↓

Confirmation

Differences between qualitative and quantitative research problem:

Qualitative research problem	*Quantitative research problem*
Explore, understands, describes, generates, discovers phenomenon	Explains or predicts the connections or comparisons between variables
Learn and describe individual view	Contains independent and dependent variable
Assess a process over time	Measures variables for getting quantified data
Generate theories based on participants view	Tests theories or broad explanations
The results cannot be generalized	Generalize results to a large number of people

Steps in formulating a research problem:

1. Identify a Research Area:

- Begin by identifying a broad area or topic of interest. Consider what field of study, subject, or theme you want to explore.

2. Review Existing Literature:

- Conduct a literature review to understand the current state of knowledge in your chosen area. This will help you identify gaps, controversies, or areas where further research is needed.

3. Ask Questions:

- Start asking questions related to your chosen area. These questions should be specific and clear. They can be exploratory in nature and should reflect your curiosity and desire to understand the subject better.

4. Define the Scope:

- Determine the scope of your research. Are you interested in a particular aspect of the topic, a specific time period, a particular location, or a particular group of people? Define the boundaries of your study to make it manageable.

5. Formulate a Research Problem Statement:

- Craft a clear and concise research problem statement. This statement should be a single sentence that summarizes the essence of your research problem. It should be specific, measurable, and address a significant issue in your field.

6. Consider the Significance:

- Reflect on why your research problem is important. How does it contribute to the existing body of knowledge? What are the potential implications and benefits of solving or addressing this problem?

7. Conduct Preliminary Research:

- Gather some initial data or information to support your research problem.

This can help you refine your problem statement and ensure that it's researchable.

8. Test Feasibility:

❖ Assess the feasibility of conducting research on the chosen problem. Consider factors such as data availability, resources, and time constraints. Ensure that your problem is realistic and can be addressed within the constraints of your study.

9. Seek Feedback:

❖ Share your research problem statement with peers, advisors, or mentors. Feedback from others can help you refine and improve your problem statement.

10. Refine and Finalize:

❖ Based on the feedback and your own reflection, refine and finalize your research problem statement. Make sure it's specific, clear, and well-defined.

11. Develop a Research Question or Hypothesis:

❖ From your research problem statement, develop specific research questions or hypotheses that you will investigate in your study. These questions or hypotheses will guide the rest of your research design and methodology.

12. Justify the Research:

❖ Write a brief rationale for why your research is necessary. Explain why your research problem is worth studying and how it contributes to the field of knowledge.

13. Provide a Working Title:

❖ Create a working title for your research project that reflects the essence of your research problem.

14. Get Approval:

❖ If you're a student or conducting research within an institution, seek approval or feedback from your advisor or research committee before proceeding.

15. Proceed with Research:

❖ Once you have a well-defined research problem, you can proceed with designing your research methodology, collecting data, and analysing results.

Strategies for writing problem statement:

❖ Clarity: Be clear and specific in describing the problem.

❖ Importance: Explain why the problem is significant.

❖ Context: Provide background and relevant context.

❖ Research Question: State the research question.

❖ Measurability: Make the problem measurable.

❖ Scope and Limitations: Define what's included and excluded.

- Originality: Address a gap or provide a fresh perspective.
- Avoid Assumptions: Base the problem on evidence.
- Testability: Ensure its amenable to empirical investigation.
- Feedback: Seek input from peers and mentors.
- Revision: Expect and be open to revisions.
- Audience: Consider your target audience.
- Structure: Use an inverted pyramid structure.
- Clarity: Write clearly and fluently.

Research purpose statement

- According to *Locke et al.* (2007), the purpose statement indicates "why you want to do the study and what you intend to accomplish"
- Research Purpose statement provides critical information about direction of a research study
- They also raise question that research will answer through the data collection procedure A good quantitative purpose statement contains:
    - The variables (independent, dependent and mediating, moderating or control)
    - The participants
    - The research sites

Writing effective purpose statements

- Specific and precise-not general, broad or obscure
- Concise – one or two sentences
- Clear – not vague, ambiguous or confusing
- Goal oriented- state in terms of desired outcomes

Guidelines for writing purpose statements

- Use a single sentence
- Wording such as *the purpose of this study ....*
- If using a theory, state the theory plan to test
- Use quantitative words (e.g., "relate" "compare" "describe") to describe the relationships between variables.

Multiple choice questions

1. In quantitative research, a common research problem is:

a) Richness of data

b) Maintaining flexibility in data collection

c) Lack of numerical data

d) Small sample size

2. **In qualitative research, which is a typical research problem?**

   a) Difficulty in applying statistical tests

   b) Lack of in-depth insights

   c) Maintaining a large sample size

   d) High cost of data analysis

3. **In quantitative research, challenges with sample selection may lead to:**

   a) Limited control over variables

   b) Poor data reliability

   c) Biased sample representation

   d) Rich descriptions

4. **A research problem in qualitative research can involve:**

   a) Hypothesis testing

   b) Exploring subjective experiences of participants

   c) Conducting large-scale surveys

   d) Statistical analysis

5. **The first step in formulating a research problem is:**

   a) Collecting data

   b) Defining the research questions

   c) Conducting a literature review

   d) Developing a research purpose statement

6. **After defining research questions, the next step in formulating a research problem is:**

   a) Collecting data

   b) Developing a research design

   c) Identifying the research purpose

   d) Selecting research methods

7. **The step that involves gathering relevant background information on the topic is:**

   a) Developing a research problem statement

   b) Identifying the research purpose

   c) Conducting a literature review

   d) Selecting research methods

8. **Which of the following is NOT a characteristic of a well-written research problem statement?**

    a) Clarity and specificity

    b) Inclusion of hypotheses

    c) Connection to the literature

    d) Feasibility

9. **The research problem statement should:**

    a) Describe the expected outcomes

    b) Define the research questions

    c) Include a detailed literature review

    d) Specify the data collection methods

10. **A research purpose statement should:**

    a) Explain how the research will be conducted

    b) Outline the expected findings

    c) Describe the research objectives and goals

    d) Present a detailed literature review

11. **Which of the following is a key element of a well-crafted research purpose statement?**

    a) Inclusion of statistical analyses

    b) A clear statement of the research's objectives

    c) A long and complex narrative

    d) A detailed review of previous research

12. **When writing a research problem statement, it is essential to:**

    a) Avoid any reference to the existing literature

    b) Keep it vague and open-ended

    c) Establish the context and significance of the problem

    d) Include technical jargon to demonstrate expertise

13. **A well-structured research purpose statement should:**

    a) Include specific data collection procedures

    b) Focus on the researcher's personal experiences and insights

    c) Offer a broad overview of the research area

    d) Omit the research objectives

14. The research problem statement and research purpose statement should be:

a) Kept separate from each other

b) Used interchangeably

c) Developed after data collection

d) Complementary and interconnected

15. In a research purpose statement, it's important to mention:

a) The personal opinions of the researcher

b) The potential limitations of the study

c) The demographic information of the participants

d) The intended contribution to the field of study

Questions

1. What challenges might researchers face in ensuring the representativeness of samples in quantitative research?
2. How do researchers in quantitative studies address issues related to maintaining data reliability and validity?
3. What are some common obstacles to exploring the rich and subjective experiences of participants in qualitative research?
4. In what ways can researcher bias impact the findings in qualitative research?
5. What challenges can small sample sizes pose in quantitative research?
6. How do researchers balance the need for structure with maintaining the flexibility of data collection in qualitative research?
7. What limitations can arise when researchers focus on numerical data and statistical analysis in quantitative research?
8. What issues may hinder the generalizability of findings in qualitative research?

References:

Jha.N.K (2008), Research methodology, research methods of qualitative and quantitative, Abhishek publications, pp no:42-54.

Ganesan, R. (2011). Research Methodologies for Engineers. MJP Publishers. Pp 9-10.

https://us.humankinetics.com/blogs/excerpt/begin-the-research-process-bychoosing-the-right-problem-or-topic-to-explore

https://www.cdsmr.com/newsworthy/qualitative-quantitative-research-the-cdsway

Chapter 8

Research Questions

Research Question

- A specific inquiry which the research seeks to provide a response to
- It resides at the core of systematic investigation and it helps us to clearly define a path for research process
- The first step in any research project
- A research question focuses on the research, determines the methodology and hypothesis and guides all stages of inquiry, analysis and reporting
- It narrows down a broad topic of interest into a specific area of study
- It influences factors such as the research methodology, sample size, data collection and data analysis

Types of Research Questions

Qualitative Research Questions:

A qualitative research question is a type of systematic inquiry that aims at collecting qualitative data from research subjects. The aim of qualitative research questions is to gather non-statistical information pertaining to the experiences, observations, and perceptions of the research subjects in line with the objectives of the investigation.

Types of Qualitative Research Questions

Ethnographic Research Questions

It is a qualitative research approach that involves observing variables in their natural environments or habitats in order to arrive at objective research outcomes. These can be used in education, business, medicine, and other fields of study, and they are very useful in contexts aimed at collecting in-depth and specific information that are peculiar to research variables.

This type of research question can be administered physically through one-onone interviews, naturalism (live and work), and participant observation methods.

Alternatively, the researcher can ask ethnographic research questions via online surveys and questionnaires.

Examples:

- Why do you use this product?
- Have you noticed any side effects since you started using this drug?

Case Studies

A case study is a qualitative research approach that involves carrying out a detailed investigation into a research subjects or variables. Here, the researcher gathers a range of data from multiple sources of information via different data collection methods, and over a period of time.

This research method can be explanatory, descriptive , or exploratory depending on the focus of your systematic investigation or research.

This type of case study uses "how" and "why" questions in order to gather valid information about the causative factors of an event.

On the other hand, exploratory case studies aim at providing answers to "who" and "what" questions using data collection tools like interviews and questionnaires.

Examples:

- Why did you choose our services?
- How has this policy affected your business output?

Interviews

An interview is a qualitative research method that involves asking respondents a series of questions in order to gather information about a research subject. They can be close-ended or open-ended, and they prompt participants to provide valid information that is useful to the research.

An interview may also be structured, semi-structured, or unstructured, and this further influences the types of questions they include. Structured interviews are made up of more close-ended questions because they aim at gathering quantitative data while unstructured interviews consist, primarily, of open-ended questions that allow the researcher to collect qualitative information from respondents.

Examples:

- What challenges did you face while using our product?
- What specific needs did our product meet?

Quantitative Research Questions

Quantitative research questions are questions that are used to gather quantifiable data from research subjects. These types of research questions are usually more specific and direct because they aim at collecting information that can be measured; that is, statistical information.

Types of Quantitative Research Questions

Descriptive Research Questions

Descriptive research questions are inquiries that researchers use to gather quantifiable data about the attributes and characteristics of research subjects. These types of questions primarily seek responses that reveal existing patterns in the nature of the research subjects. They are typically closed-ended because they aim at gathering definite and specific responses from research participants.

Examples

- How often do you make use of our fitness application?
- How much would you be willing to pay for this product?

Comparative Research Questions

A comparative research question is a type of quantitative research question that is used to gather information about the differences between two or more research subjects across different variables. These types of questions help the researcher to identify distinct features that mark one research subject from the other while highlighting existing similarities.

The 5 steps involved in the framing of comparative research questions are:

- Choose your starting phrase
- Identify and name the dependent variable
- Identify the groups you are interested in
- Identify the appropriate adjoining text
- Write out the comparative research question

Examples:

- What are the differences between a landline telephone and a smartphone?
- What are the differences between work-from-home and on-site operations?

Relationship-Based Research Questions

A relationship-based research question is one that inquires into the nature of the association between two research subjects within the same demographic. These types of research questions help you to gather information pertaining to the nature of the association between two research variables.

Relationship-based research questions are also known as correlational research questions because they seek to clearly identify the link between 2 variables.

Examples:

- What is the relationship between purchasing power and the business site?
- What is the relationship between the work environment and workforce turnover?

Good Research Questions:

A good research question is one that requires impartial responses and can be answered via existing sources of information.

1. Open-Ended Questions

Anopen-ended question is a type of research question that does not restrict respondents to a set of premeditated answer options. In other words, it is a question that allows the respondent to freely express his or her perceptions and feelings towards the research subject.

Examples:

- ❖ How do you deal with stress in the workplace?
- ❖ What is a typical day at work like for you?

2. Close-ended Questions

Aclose-ended question is a type of survey question that restricts respondents to a set of predetermined answers such asmultiple-choice questions. Close-ended questions typically require yes or no answers and are commonly used in quantitative research to gather numerical data from research participants.

Examples:

- ❖ Did you enjoy this event?

  Yes

  No

3. Likert Scale Questions

A Likert scale question is a type of close-ended question that is structured asa 3point, 5-point, or 7-point psychometric scale. This type of question is used to measure the survey respondent's disposition towards multiple variables and it can be unipolar or bipolar in nature.

Examples:

- ❖ How satisfied are you with our service delivery?

1. Very dissatisfied
2. Not satisfied
3. Neutral
4. Satisfied
5. Very satisfied

4. Rating Scale Questions

Arating scalequestion is a type of close-ended question that seeks to associate a specific qualitative measure (rating) with the different variables in research.

It is commonly used in customer experience surveys, market research surveys, employee reviews, and product evaluations.

Examples:

- How would you rate our service delivery?
  1. Excellent
  2. Good
  3. Neutral
  4. Bad
  5. Very bad

Bad Research Question

They help you avoid them in the course of your systematic investigation. These types of questions are usually unfocused and often result in research biases that can negatively impact the outcomes of your systematic investigation.

1. Loaded Questions

A loaded question is a question that subtly presupposes one or more unverified assumptions about the research subject or participant.

Examples:

- Have you stopped smoking?
- Where did you hide the money?

3. Negative Questions

A negative question is a type of question that is structured with an implicit or explicit negator. Negative questions can be misleading because they upturn the typical yes/no response order by requiring a negative answer for affirmation and an affirmative answer for negation.

Examples:

- Would you mind dropping by my office later today?
- Didn't you visit last week?

4. Leading Questions

A leading questionis a type of survey question that nudges the respondent towards an already-determined answer. It is highly suggestive in nature and typically consists of biases and unverified assumptions that point towards its premeditated responses.

Examples:

- If you enjoyed this service, would you be willing to try out our other packages?
- Our product met your needs, didn't it?

Criteria for Selecting Research Question

- **Focused** on a single problem or issue
- **Researchable** using primary and/or secondary sources
- **Feasible** to answer within the timeframe and practical constraints
- **Specific** enough to answer thoroughly
- **Complex** enough to develop the answer over the space of a paper or thesis
- **Relevant** to your field of study and/or society more broadly

Techniques for Narrowing a Problem into a Research Question

After choosing a topic and gathering background information, add focus with a research question

Explore Questions

- Ask open ended 'how' and 'why' questions about your problem
- Consider the 'so what' of your topic
- Reflect on the questions you have considered. Identify one or two questions that you find engaging and which could be explored through research

Determine and Evaluate your Research Question

- what aspect of the general topic you will explore
- is your research question clear
- is your research question focused
- is your research question complex (it shouldn't have simple yes/no answer and should require research and analysis)

Hypothesize

- After you have come up with a question, consider the path your answer might take
- If you are making an argument, what will you say?
- Why does your argument matter?
- How might others challenge your argument?
- What kind of sources will you need to support your argument?

References

C.R. Kothari,1985, Research Methodology: Methods and Techniques

Pujari Daivadeenam and Laxmi Lal Somani – Research Methodology in Extension Education , 2nd Edition

A K Singh, 1986,Tests,Measurements and Research methods in Behavioural Sciences

Fred N Kerlinger,2017,Foundations of behavioural Research

https://research.com>how to write a research question – types, steps

https://www.google.com/search?q=types+of+research+questions+pdf

Questions

- What is research question?
- How many types there in research questions?
- What is qualitative research question?
- What is mean by quantitative research question?
- Open-ended question comes under which type of research question?
- Give me a definition for relationship based research question?
- What all are the criteria is need for selecting research question?
- What all are the techniques are there to narrowing a problem into research question?
- What is mean by hypothesize?
- Definition – Ethnographic research question.

Fill In the Blanks

- A search questions should be__________and__________
- A __________research question aims to understand the experiences, perceptions or meanings of individuals
- When selecting a research questions its important to consider ________ and__________feasibility.
- Research questions are _________that guide the direction of research study
- __________research question seeks to explore relationships or differences between variables.
- The two main types of research questions are _______and_________
- When selecting research questions its crucial to consider the _______of the study
- An__________ is a type of research question that does not restrict respondents to a set of premeditated answer options. In other words, it is a question that allows the respondent to freely express his or her perceptions and feelings towards the research subject.
- An interview is a______ research method.
- A rating scale question is a type of_____questions.

Chapter 9

Objectives: Meaning, Types and Criteria

Research methodology plays a crucial role in the field of agricultural extension. It provides the framework for conducting effective research that can inform policy, improve practices, and enhance the overall agricultural sector. A fundamental aspect of research methodology is setting clear objectives for a study. In this comprehensive discussion, we will delve into the meaning of objectives, explore the different types of objectives, examine the criteria for judging objectives, and understand the importance of crafting a well-structured statement of objectives in the context of agricultural extension research.

Introduction to Objectives

Objectives are the backbone of any research endeavor, serving as the guiding lights that illuminate the path towards achieving the research goals. They are essential components of a research study as they provide clarity and purpose to the entire research process. Simply put, objectives help researchers answer the fundamental question: What are we trying to accomplish with this research?

Meaning of Objectives

In the context of agricultural extension research, objectives are clear, concise, and specific statements that delineate the goals and intentions of a research project. They are the 'what' of research, defining precisely what the researcher aims to accomplish. Objectives serve as a compass, ensuring that the research remains focused and does not drift off course.

Objectives should be devoid of ambiguity and vagueness. They should be explicit in articulating the research's purpose and direction. When crafting objectives, it's vital to remember that they are the foundation upon which the entire research process is built. Therefore, precision is key.

Definitions of Objective

Mc. Farland defines objectives, "Objectives are the goals, aims, or purposes that organizations wish to achieve over varying periods of time."

In the words of **Terry**, "A managerial objective is the intended goal which prescribes definite scope and suggests direction to efforts of a manager."

Objectives are expressions of the ends towards which our efforts are directed. **--Anonymous**

Objective is a statement of change in knowledge, attitude and skill which we want to study among the people. – **Anonymous**

e.g. Change in knowledge: Too heavy application of nitrogen leads to excessive lodging of paddy crop.

Change in Attitude: There is no sin in killing rats which are responsible for huge national loss of foodstuffs.

Change in skills: How to use zinc phosphide as a poison bait without undue risk-change in skill.

Types of Objectives

There are three levels or types of objectives:

(i) Fundamental objectives

(ii) General objectives

(iii) Working objective

(i) Fundamental objectives: They are all inclusive objectives. The fundamental objective of extension work is the development of the people or the Destination of Man. M.C. Burritt, a master farmer and former. Director of extension, New York state said. It is the function of the Extension Service to teach people to determine accurately their own problems, to help them to acquire knowledge and to inspire them to action, but it must be their own action out of their own knowledge and convictions.

The fundamental objective of extension is to develop the rural people economically, socially and culturally by means of education. — Anonymous

e.g. The good life, better citizenship, development of individual.

(ii) General objectives: These are more definite than fundamental objectives, and are directly associated with the extension service and are generally found in statements of policies and purposes.

e.g. Helping rural people to have better home living.

More specifically stated, the following are general objectives of Extension:

a. To assist people to discover and analyze their problems, their felt and unfelt needs.
b. To develop leadership among people and help them in organizing groups, to solve their problems.

c. To disseminate information based on research and / or practical experience, in such a manner that the people would accept it and put it into actual practice.

d. To keep the research workers informed of the people's problems from time to time, so that they may offer solutions based on necessary research.

(iii) Working objectives: They have specific subject matter approach. Working objective have three parts.

a) Audience who are to be changed

b) Type of change-knowledge, attitude or skill

c) Content or message-What you are trying to teach (What you want them to know, feel or do)?

e.g. The maize growers of village X to increase the yield per acre of maize by using hybrid seed.

The major objectives of extension may also be categorized as follows:

a) Material-increase production, income

b) Educational-change the outlook of people or develop the individuals

c) Social and cultural-development of the community.

Criteria for Judging Objectives

Setting objectives is not a trivial task; it requires careful consideration to ensure that they are well-crafted and effective. To assess the quality of objectives, researchers often use the SMART criteria, which stands for Specific, Measurable, Achievable, Relevant, and Time-bound:

Are they specific, consistent and worded clearly?

Objectives should be clear, precise, and unambiguous. They should leave no room for interpretation or confusion. Specific objectives provide a distinct focus for the research.

Can they be measured (or) evaluated?

Objectives should be formulated in a way that allows for measurement or evaluation of progress. This means that there should be concrete criteria or indicators to determine whether the objectives have been achieved.

Are they achievable by the level of maturity of group and permitted by the available resources?

Objectives should be realistic and attainable within the resources, time frame, and constraints of the research project. Setting objectives that are too ambitious can lead to frustration and failure.

Are they relevant?

Objectives should be pertinent to the research problem and align with the broader research goals. They should directly contribute to addressing the research questions or issues at hand.

Can they be achieved-with limited time at the disposal of the investigator?

Objectives should have a clear time frame or deadline for completion. This adds a sense of urgency and helps in managing the research project's timeline.

Are they limited in number, to avoid undue confusion and diffusion of efforts?

By adhering to the SMART criteria, researchers can ensure that their objectives are well-defined, actionable, and conducive to successful research outcomes.

Importance of Clear Objectives

Clear objectives are paramount in agricultural extension research for several reasons:

Guiding Research: Objectives serve as a guiding light throughout the research process. They help researchers stay on track and maintain focus on the research goals.

Communication: Clear objectives facilitate effective communication of research intentions to stakeholders, collaborators, and funders. They ensure that everyone involved in the research understands its purpose and direction.

Evaluation: Objectives provide a basis for evaluating the success of the research project. Researchers can measure their progress against the stated objectives to determine whether they have achieved their goals.

Decision-Making: Objectives guide decision-making at various stages of the research, from research design and data collection to data analysis and interpretation. They help in prioritizing research activities.

Statement of Objectives

A well-structured statement of objectives is an essential component of any research project. It encapsulates the essence of the research by providing a concise summary of the general and specific objectives, along with any operational objectives if applicable. A properly crafted statement of objectives sets the stage for the research and helps in gaining a clear understanding of what the study aims to achieve.

A typical statement of objectives in agricultural extension research includes the following elements:

A clear and concise description of the general objective. This is the overarching goal of the research and should be stated in a way that encapsulates the research's primary purpose.

A list of specific objectives or research questions. Each specific objective should be outlined in a clear and straightforward manner, reflecting the narrower goals that contribute to the general objective.

If applicable, any operational objectives may be included. Operational objectives should be specific and practical, outlining the steps or tasks required to achieve the specific objectives.

Here's an example of a statement of objectives for an agricultural extension research project:

Statement of Objectives:

1. General Objective:

"To improve the adoption of sustainable farming practices among small-scale farmers."

2. Specific Objectives:

"To assess the current level of adoption of sustainable farming practices among small-scale farmers in the study area."

"To identify the factors influencing the adoption of sustainable farming practices."

"To develop and test a training program aimed at promoting sustainable farming practices among small-scale farmers."

3. Operational Objectives (if applicable):

"To survey 200 small-scale farmers in the target area using a structured questionnaire."

"To conduct focus group discussions with 10 groups of farmers to gather qualitative insights."

"To deliver training workshops to 50 farmers and assess the impact on their farming practices."

This statement of objectives provides a comprehensive overview of the research project, from its broad goals to the specific tasks that will be undertaken to achieve those goals.

Conclusion

In conclusion, objectives are foundational in agricultural extension research, categorized as general, specific, and operational. Adhering to SMART criteria ensures their effectiveness, guiding research, communication, evaluation, and decision-making. Well-crafted objectives are vital, especially in agricultural extension, aiming to enhance farming practices and sustainability. By setting clear objectives, researchers drive impactful change, making them essential in agricultural research and development.

Reference

Nakkiran, S., G. Ramesh (2009). Research Methods in Rural development. Deep & Deep Publications. Pp

Ganesan, R. (2011). Research Methodologies for Engineers. MJP Publishers. Pp 5-6

Daivadeenam, P., L.L. Somani. (2002). Research Methodology in Extension Education. 2nd Edition. ATPA. Pp 108-113

Multiple Choice Questions

1) What role do objectives play in agricultural extension research?

a) They are optional components of a research project.

b) They provide clarity and purpose to the research process.

c) They are only relevant to theoretical research.

d) They are primarily used for data analysis.

2) Which term best defines objectives in the context of agricultural extension research?

a) Hypotheses

b) Goals

c) Outcomes

d) Tools

3) What is the main purpose of objectives in research?

a) To confuse researchers

b) To provide a foundation for research

c) To generate multiple research questions

d) To create ambiguity

4) According to the SMART criteria, what does "M" stand for in the context of objectives?

a) Mysterious

b) Measurable

c) Mandatory

d) Motivated

5) Which level of objectives is the most fundamental and all-inclusive?

a) General objectives

b) Working objectives

c) Fundamental objectives

d) Abstract objectives

6) Fundamental objectives in agricultural extension research are primarily concerned with:

a) Policy implementation

b) Developing individuals and communities

c) Resource allocation

d) Research design

7) General objectives in agricultural extension research are more:

a) Detailed and specific

b) Ambiguous and vague

c) Academic in nature

d) Focused on long-term outcomes

8) What is the primary focus of working objectives in research?

a) Organizational development

b) Identifying research problems

c) Concrete subject matter and specific actions

d) Setting long-term goals

9) When setting objectives, why is it important to make them specific?

a) To create ambiguity

b) To leave room for interpretation

c) To provide a distinct focus for research

d) To confuse stakeholders

10) What is the significance of ensuring that objectives are measurable or evaluable?

a) It simplifies the research process.

b) It allows for tracking progress and success.

c) It increases ambiguity in research.

d) It hinders collaboration.

11) Objectives should be achievable by the level of maturity of the group and permitted by the available resources. This implies that objectives should be:

a) Unrealistic and ambitious

b) Realistic and attainable

c) Vague and unclear

d) Ignorant of available resources

12) How do objectives contribute to decision-making in research?

a) They have no impact on decision-making.
b) They prioritize research activities.
c) They hinder the decision-making process.
d) They create confusion.

13) What is the primary role of a statement of objectives in research?

a) To provide detailed research results
b) To describe the research process
c) To set the stage for the research and convey its goals
d) To evaluate the success of the research

14) In a statement of objectives for agricultural extension research, what does the general objective represent?

a) The specific research questions
b) The broader research goals
c) The operational objectives
d) The financial budget

15) Which element of a statement of objectives outlines the narrower goals that contribute to the general objective?

a) The general objective
b) The specific objectives
c) The operational objectives
d) The research design

16) If you need to conduct a survey of 200 small-scale farmers in a research project, which part of the statement of objectives does this detail fall under?

a) General Objective
b) Specific Objectives
c) Operational Objectives
d) Research Design

17) Which of the following is NOT one of the SMART criteria for judging objectives?

a) Specific
b) Sustainable
c) Measurable
d) Relevant

18) Clear objectives are important in research for all of the following reasons EXCEPT:

a) Guiding the research process

b) Facilitating effective communication

c) Creating ambiguity

d) Aiding decision-making

19) A well-structured statement of objectives helps in gaining a clear understanding of what the study aims to achieve. What is the term for this purpose?

a) Research statement

b) Research methodology

c) Research proposal

d) Research objectives

20) In agricultural extension research, why is it essential to have clear objectives in the research process?

a) To confuse stakeholders

b) To hinder communication

c) To maintain focus on research goals

d) To increase ambiguity in research

Answers

1. b) They provide clarity and purpose to the research process.
2. b) Goals
3. b) To provide a foundation for research
4. b) Measurable
5. c) Fundamental objectives
6. b) Developing individuals and communities
7. a) Detailed and specific
8. c) Concrete subject matter and specific actions
9. c) To provide a distinct focus for research
10. b) It allows for tracking progress and success.
11. b) Realistic and attainable
12. b) They prioritize research activities.
13. c) To set the stage for the research and convey its goals
14. b) The broader research goals

15. b) The specific objectives
16. c) Operational Objectives
17. b) Sustainable
18. c) Creating ambiguity
19. d) Research objectives
20. c) To maintain focus on research goals

Fill in the Blank Question

1. Objectives in agricultural extension research serve as the guiding lights that illuminate the path towards achieving the research ________.
2. In agricultural extension research, objectives should be clear, concise, and specific statements that delineate the ________ and intentions of a research project.
3. The foundation of the entire research process is built upon well-crafted objectives, making ________ a key element.
4. Mc. Farland defines objectives as the goals, aims, or purposes that organizations wish to achieve over varying periods of ________.
5. Terry describes managerial objectives as intended goals that prescribe a definite scope and suggest ________ to a manager's efforts.
6. SMART criteria are often used to assess the quality of objectives, where SMART stands for Specific, Measurable, Achievable, Relevant, and ________.
7. Objectives should be formulated in a way that allows for ________ or evaluation of progress.
8. Setting objectives that are too ambitious and not achievable within the available resources can lead to ________ and failure.
9. Clear objectives facilitate effective ________ of research intentions to stakeholders, collaborators, and funders.
10. A well-structured statement of objectives in agricultural extension research provides a concise summary of the general and specific objectives, along with any ________ objectives if applicable.

Answers

1. Goals
2. Goals
3. Precision
4. Time
5. Direction

6. Bound
7. Measurement
8. Frustration
9. Communication
10. Operational

Questions

1. What is the definition of objective in research?
2. How many types of objectives exist in research, and what are they?
3. What is a fundamental objective?
4. What is a general objective?
5. What is a working objective?
6. What are the criteria for judging objectives?
7. What is the importance of objectives in research?

Chapter 10

Literature Review

Literature Review

A literature review is a critical and evaluative description of publications on a chosen topic. Literature reviews are the integral parts of research proposals, theses and research papers. Review of literature is also helpful in developing expertise, identifying the contribution that can be made and developing appropriate research methodology.

Objectives of Literature Review

- To understand the historical background of the research
- To view current context of the research referring to contemporary debates on issues and questions
- To explore relevant theories and concepts that underpin the research
- To know relevant terminology, their definitions and context in which the terms are used
- To understand related literature and know how the research issues have been addressed

Importance of Literature Review

1. Demonstrate a Clear Understanding of the Subject

Writing a literature review demonstrates that you have a clear understanding of the subject you're investigating. It also means that you can easily identify, evaluate, and summarize existing research that's relevant to your work.

2. Justify Your Research

There's more to writing a research paper than just identifying topic and generating your research question from it. You also have to go as far as to justify your research, and the only way to do that is by including a literature review in your work. It's important to understand that looking at past research is the only way to identify gaps that exist in the current literature. That can go a long way to help fill in the gap by addressing them in your own research work.

3. Helps to Set a Resourceful Theoretical Framework

Because a research paper assignment builds up on the ideas of already existing research, doing a literature review can help you to set a resourceful theoretical framework on which to base your study. The theoretical framework will include concepts and theories that you will base your research on. And keep in mind that it's this framework that professors will use to judge the overall quality of your work.

Functions of Literature Review

There are five functions of review of literature:

- The conceptual frame of reference for the contemplated research.
- An understanding of the status of research in problem area.
- Clues to the research approach, method, instrumentation and data analysis.
- An estimate of the probability of success of the contemplated research and the significance or usefulness of the findings and, assuming the decision is made to continue and
- Specific information required to interpret the definitions, assumptions, limitations and hypotheses of research.

Purpose of literature review

The purpose of a literature review is to collect relevant, timely research on your chosen topic, and synthesize it into a cohesive summary of existing knowledge in the field. This then prepares you for making your own argument on that topic, or for conducting your own original research.

Steps involved in conducting literature review

There are four steps involved in conducting literature review;

- Search for existing literature in your area of study.
- Review the literature selected.
- Develop a theoretical framework.
- Develop a conceptual framework.

Search for existing literature in your area of study

- To effectively search for literature in your field of inquiry, it is imperative that you have in mind at least some idea of the broad subject area and of problem you wish to investigate, in order to set a parameters for your search
- Compile a bibliography for this broad area using books and journals as sources
- The best way to search for a book is to look at your library catalogues and publications such as Book review index can help you to locate the books you needed
- There are several sources like indices of journals, abstracts of articles and citation indices make your search for journals easier

Review the literature selected

- Note whether the knowledge relevant to your theoretical framework has been confirmed beyond doubt
- Note the theories put forward, the criticisms of these and their basis, the methodologies adopted (study design, sample size and its characteristics, measurement procedures, etc.) and the criticisms of them
- Examine to what extent the findings can be generalised to other situations
- Notice where there are significant differences of opinion among researchers and give your opinion about the validity of these differences
- Ascertain the areas in which little or nothing is known – the gaps that exist in the body of knowledge

Developing a theoretical framework

- The information obtained from different books and journals now needs to be sorted under the main themes and theories, highlighting agreements and disagreements among the authors and identifying the unanswered questions or gaps
- You will also realise that the literature deals with a number of aspects that have a direct or indirect bearing on your research topic
- Use these aspects as a basis for developing your theoretical framework
- Your review of the literature should sort out the information, as mentioned earlier, within this framework
- Read some of the literature and then attempt to develop a framework, even a loose one, within which you can organise the rest of the literature you read
- As you read more about the area, you are likely to change the framework
- However, without it, you will get bogged down in a great deal of unnecessary reading and note-taking that may not be relevant to your study
- Literature pertinent to your study may deal with two types of information: 1. universal and 2. more specific (i.e. local trends or a specific programme)

Developing a conceptual framework

- The conceptual framework is the basis of your research problem. It stems from the theoretical framework and usually focuses on the sections which become the basis of your study
- Whereas the theoretical framework consists of the theories or issues in which your study is embedded, the conceptual framework describes the aspects you selected from the theoretical framework to become the basis of your enquiry the conceptual framework is focused on indicators to measure the success or failure of the strategies to enhance community responsiveness
- Hence the conceptual framework grows out of the theoretical framework and relates to the specific research problem

Identifying key terms

- It is important to find all the relevant key terms for the topic to ensure the search is comprehensive by identifying different spellings, tenses and word variants of keywords, concepts and name of the author associated with your topics

There are many ways to locate these terms, including

- Recommended readings, textbooks and other review articles that provide an overview of the field of research
- Dictionaries, handbooks and encyclopedias that provide definitions and general information about topics
- Database or subject headings that tell you which terms are used in the databases and professional literature
- Text mining tools that allow you to analyse large amounts of text or information and identify commonly used terms in the field

Locate literature review

- One of the essential preliminary tasks when you undertake a research study is to go through the existing literature in order to acquaint yourself with the available body of knowledge in your area of interest
- Reviewing the literature can be time consuming, daunting and frustrating, but it is also rewarding. The literature review is an integral part of the research process and makes a valuable contribution to almost every operational step
- It has value even before the first step; that is, when you are merely thinking about a research question that you may want to find answers to through your research journey
- In the initial stages of research it helps you to establish the theoretical roots of your study, clarify your ideas and develop your research methodology
- Later in the process, the literature review serves to enhance and consolidate your own knowledge base and helps you to integrate your findings with the existing body of knowledge
- Since an important responsibility in research is to compare your findings with those of others, it is here that the literature review plays an extremely important role
- During the write-up of your report it helps you to integrate your findings with existing knowledge – that is, to either support or contradict earlier research
- The higher the academic level of your research, the more important a thorough integration of your findings with existing literature becomes

Locating relevant literature

- Information available in a number of formats can be accessed from different sources
- However, it is important to understand the significance of various formats so that researchers know what is more suited to their information requirements
- When you plan, you should aim for a comprehensive literature search

Select Appropriate Tools

- There are several methods for locating relevant literature
- These include searching through library catalogues, abstracts, indexes, general subject guides, CD-ROM databases, search engines, online databases such as CAB Direct and PubMed. checking library links on the web, and consulting many other Internet resources

Trace Relevant Publications

- Using the selected tools, search through the subject indexes to trace relevant articles. You may use appropriate key words for an effective search
- Use synonyms and alternative terms of key words also. Try to find out as many reference sources as possible on the topic
- However, if you find too many articles, use keywords that are more specific. If you are not able to find appreciable number of references, you have to change your search strategies
- Broaden your search using more general terms. Still, if you are not able to trace out much information, the reason may be that you are using an inappropriate index or there are not many previous works on it
- Sometimes, the topic selected may be so new that there has been no time for published research to appear in print

Note Taking and Note Making

- Although the two phrases note taking and note making are often used interchangeably, they are not the same
- Note taking is a passive process, mostly done by the students or listeners during lectures, but note making is a more active and focused activity where you understand the information, and note down only what is relevant and important for you
- In other words, note taking is almost verbatim recording from a book or lecture, and note making is a summary representation or comment in your own words on what you read, heard
- Make good notes including the details of all relevant sources you found and read

Methods of Note taking and Note making

Listed below are a few methods of note-making.

- The Outlining Method.
- The Mapping Method.
- The Sentence Method.
- The Cornell Method.
- The Charting Method.

Systems of Note-making

The utilisation of bullet points is something we generally run over in different books. The fundamental thought behind this application is giving substance about something, in a manner that isn't effectively forgettable. The reason for note-making is something similar; in bringing the attention of perusers to the main data, to pull back beyond all the other things.

1. Heading and Subheading:

To assign a short and best-got heading and subheading, one must go through the source content to some extent twice.

2. Content:

Just compose the most pertinent substance to make your notes genuinely exact. Furthermore, remember to incorporate those significant subtleties that you figure you will struggle reviewing later.

3. Comprehensibility:

There are a couple of tips that one can use to stay away from any uncertainty to make the notes more snappy. Try not to pass on pointless spaces to keep away from disarray. Utilise however many acronyms as it would be prudent, yet, it ought not to come out as a helpless understanding.

4. Formats of Note-making:

A note-making design includes the choosing, examination, outlining, and association of data. There are distinctive note-production designs. Allow us to make you acquainted with them. Graph or Pattern design (A visual note-making design strategy): In this technique, the data is presented by a chart. Related thoughts associate the fundamental subject. This technique is outwardly more clear.

Mind Map (A visual note-making strategy):

Here, the central matters and all the related focuses are introduced through a guide or a map. It contains text and pictures. For an unmistakable agreement, the data is connected in the appropriate grouping.

Layout or Outline Format (A visual note making design technique):

Here, the data is introduced as a blueprint. Appropriate titles and captions are numbered in like manner for this blueprint. Question and Prompt Format (A visual strategy): Here, the central matters are featured as a progression of inquiries with suitable replies. It helps in rapidly going through the notes.

Split Page Format (A descriptive form of note-making technique):

In this strategy, the page is partitioned into two sections. The primary section has the standard notes. The subsequent segment will have a rundown or a summary. This is the most widely recognised arrangement for jotting down in a meeting or lecture notes.

Critical evaluation and selection

The material collected from different sources may vary in quality, completeness and accuracy. The opinion expressed and conclusions drawn at even under similar situations may be different from one source to the other or at times may be conflicting and contradictory. Evaluation of literature should be therefore, done considering the following aspects:

- Reliability and dependability of the source
- Updated information
- Suitability for meeting the specific requirement of research
- Scope of data
- Treatment of subject matter and style
- Arrangement of subject matter
- Illustrative tables and diagrams
- Target group of users, etc.

A researcher should accept the data only after specified requirements are fulfilled. The collected material should be organized first chapter wise and then within each chapter item-wise, so that the same may be recalled and used for writing reviews of literature as well as while writing research report.

Organizing literature

A literature review is a piece of discursive prose, not a list describing or summarizing one piece of literature after another. It should have a single organizing principle

Chronological

- The chronological framework organizes the literature in the order in which they are published
- For example, if you were writing about a specific teaching method, you would begin with the materials that first introduced the method

- You would then follow with case studies applying that method
- You would conclude your review with contemporary papers that may even give a historical perspective on the method from when it was first conceived and how it is applied today

Thematic

- The thematic framework is similar to organizing by trend, except that you are not organizing the reviews in the order that they were published
- This does not mean that you do not consider the timeline for how a topic or issue developed, but that you will not focus on organizing your reviews chronologically
- Rather, the emphasis will be on the themes you find within the topic or issue such as commonalities and from there you fit your reviews into the separate ideas in which they fit
- For example, if the review topic was arts-based research, your review may focus on different ways artistic inquiry was used to understand the creative process, focusing then on the concepts rather than the development

Methodological

- The method or practice applied in a case study can be the basis for organizing a literature review
- This framework focuses on how the authors or the persons administering a study applied similar methods as another study
- As a result, the types of literature in a literature review that applies the methodological framework tends to review similar material

5 C's of writing a literature review

- **Cite** the material you have referred to and used to help you define the research problem that you will study
- **Compare** the various arguments, theories, methods, and findings expressed in the literature
- **Contrast** the various arguments, themes, methods, approaches, and controversies apparent and/or described in the literature
- **Critique** the literature, describe which arguments you find more persuasive and explain why, explain which approaches, findings, and methods seem most reliable, valid, appropriate, and/or most popular and why pay attention to the verbs you use to describe what previous researchers have stated
- **Connect** the various research studies you reviewed, describe how your work utilizes, draws upon, departs from, synthesizes, adds to or extends previous research studies

Benefits of Conducting Literature Review

1. Assessment of the current state of research on a topic.

This is probably the most obvious value of the literature review. Once a researcher has determined an area to work with for a research project, a search of relevant information sources will help determine what is already known about the topic and how extensively the topic has already been researched.

2. Identification of the experts on a particular topic.

One of the additional benefits derived from doing the literature review is that it will quickly reveal which researchers have written the most on a particular topic and are, therefore, probably the experts on the topic. Someone who has written twenty articles on a topic or on related topics is more than likely more knowledgeable than someone who has written a single article. This same writer will likely turn up as a reference in most of the other articles written on the same topic. From the number of articles written by the author and the number of times the writer has been cited by other authors, a researcher will be able to assume that the particular author is an expert in the area and, thus, a key resource for consultation in the current research to be taken.

3. Identification of key questions about a topic that need further research.

In many cases a researcher may discover new angles that need further exploration by reviewing what has already been written on a topic. For example, research may suggest that listening to music while studying might lead to better retention of ideas, but the research might not have assessed whether a particular style of music is more beneficial than another. A researcher who is interested in pursuing this topic would then do well to follow up existing studies with a new study, based on previous research, that tries to identify which styles of music are most beneficial to retention.

4. Determination of methodologies used in past studies of the same or similar topics.

It is often useful to review the types of studies that previous researchers have launched as a means of determining what approaches might be of most benefit in further developing a topic. By the same token, a review of previously conducted studies might lend itself to researchers determining a new angle for approaching research.

Some of the guidelines for conducting literature review

- Include an introduction and conclusion. Even if the literature review will be part of a longer document, introductory and concluding paragraphs can act as bookends to your material. Provide background information for your reader, such as including references to the pioneers in the field in the beginning and offering closure in the end by discussing the implications of future research to the field.

- ❖ Avoid direct quotations. Just like in an annotated bibliography, you will want to paraphrase all of the material you present in a literature review. This assignment is a chance for you to demonstrate your knowledge on a topic, and putting ideas into your own words will ensure that you are interpreting the found material for your reader. Paraphrasing will also ensure your review of literature is in your authorial voice.
- ❖ Organize by topic or theme rather than by author. When compiling multiple sources, a tendency can be to summarize each source and then compare and contrast the sources at the end. Instead, organize your source information by your identified themes and patterns. This organization helps demonstrate your synthesis of the material and inhibits you from creating a series of book reports.
- ❖ Use headings. APA encourages the use of headings within longer pieces of text to display a shift in topic and create a visual break for the reader. Headings in a literature review can also help you as the writer organize your material by theme and note any layers, or subtopics, within the field.
- ❖ Show relationships and consider the flow of ideas. A literature review can be lengthy and dense, so you will want to make your text appealing to your reader. Transitions and comparison terms will allow you to demonstrate where authors agree or disagree on a topic and highlight your interpretation of the literature.

References

George Thomas, C. (2021). Research Methodology and Scientific Writing. Springer Nature.

Daivadeenam, P., & Somani, L.L. (2013). Research Methodology in Extension Education (Second Edition). ATPA.

Ranjit Kumar. (2005). Research Methodology: A Step by Step Guide for Beginers (Second Edition). Pearson India

https://pressbooks.bccampus.ca/researchmethods/chapter/the-five-cs-of-writing-a-literature-review/

Answer the following

1. What are the steps involved in literature review?
2. Write the importance of literature review.
3. Quote the objectives of literature review.

4. List out the note taking methods.
5. How will you locate the key terms for literature review?
6. How will you development a conceptual framework for your literature review?
7. Define literature review.
8. Define note taking.
9. What are the 5 C's of writing literature review?
10. What are the systems of note making?
11. Difference between note taking and note making.

Fill in the blanks

1. Dictionaries, ________ and ________ that provides definition and general topics about the research.
2. Organizing literature review should have _______ principle.
3. The ________ is the basis of your research problem.
4. Note taking is a _______ process.
5. The theoretical framework consists of ________ embedded your study.
6. A researcher should accept the data only after ________ are fulfilled.

Chapter 11

Theoretical, Conceptual and Empirical Frameworks

Introduction

Definitions

- **Theoretical framework** is the structure that hold or support a theory of research study. It introduces and describes the theory that explains why research problem under study exists.
- **Conceptual framework** refers to the order and arrangement of different concepts to build a theory and postulate.
- **Empirical framework** is based on observed and measured phenomena and derives knowledge from actual experience rather than theory or belief.

Developing Theoretical Framework

1. Identify your key concepts

- To pick out the key terms from problem statement and research questions.
- Concept often have multiple definitions, so the theoretical framework involves clearly defining what you mean by each term

2. Evaluate and explain relevant theories

- Thorough literature review, you can determine how other researchers have defined and drawn connections between these key concepts.
- Aim to compare and critically evaluate the approaches that different authors have proposed.

3. Show how your research fits in

Theoretical framework should show how your own project will make use of these ideas.

- Test whether a theory holds in a specific context
- Use theory as a basis for interpreting your results
- Critique or challenge a theory

- ❖ Combine different theories in a new or unique way

If relevant, you can also use the theoretical framework to develop hypotheses for your research.

Developing Conceptual Framework

1. Choose the topic

- ❖ Decide on what will be your research topic.
- ❖ The topic should be within your field of specialization.
- ❖ Generate your research topic using brainstorming tips.

2. Choose the title

A title should focus on study and illustrate the relationship between variables.

3. Isolate the key concepts / constructs in the title

- ❖ Identify the specific variables described in the literature and determine how these are related.
- ❖ Some research abstracts contain the variables, and the salient findings thus may serve the purpose.
- ❖ If these are not available, find the research paper's summary.

4. Do a literature review

- ❖ Review relevant research on the theme that you decide to work on after scrutiny of the issue at hand.
- ❖ Usepeer - reviewed and well-known scientific journals as these are reliable sources of information.

5. Generate the conceptual framework

- ❖ The study's problem statement or research objective provide as a reference for constructing the conceptual framework.
- ❖ Your study will attempt to answer the question that other researchers have not explained yet.
- ❖ Your research should address a knowledge gap.

Developing Empirical Framework

1.Define the purpose of research

The researcher has to answer questions like

- ❖ What is the problem statement?
- ❖ What exactly do I want to find out?
- ❖ Are there any issues in terms of the availability of knowledge, data, time or resources. Will this research be more beneficial than what it will cost

- Before going ahead, a researcher has to clearly define his purpose for the research and set up a plan to carry out further tasks.

2. Supporting theories and relevant literature

- The researcher needs to find out if there are theories which can be linked to his research problem.
- He has to figure out if any theory can help him support his findings.
- All kind of relevant literature will help the researcher to find if there are others who have researched this before, or what are the problems faced during this research.
- The researcher will also have to set up assumptions and also find out if there is any history regarding his research problem

3. Creation of Hypothesis and measurement

- Before beginning the actual research he needs to provide himself a working hypothesis or guess what will be the probable result.
- Researcher has to set up variables, decide the environment for the research and find out how can he relate between the variables.
- Researcher will also need to define the units of measurements, tolerable degree for errors, and find out if the measurement chosen will be acceptable by others.

4. Methodology, research design and data collection

- The researcher has to define a strategy for conducting search.
- Set up experiments to collect data which will enable him to propose the hypothesis.
- The researcher will decide whether he will need experimental or non-experimental method for conducting the research.
- The type of research design will vary depending on the field in which the research is being conducted.
- The researcher will have to find out parameters that will affect the validity of the research design.
- Data collection will need to be done by choosing appropriate samples depending on the research question.
- To carry out the research, he can use one of the many sampling techniques.
- Once data collection is complete, researcher will have empirical data which needs to be analysed.

5. Data Analysis and result

- Data analysis can be done in two ways, qualitatively and quantitatively.
- Researcher will need to find out what qualitative method or quantitative method will be needed or will he need a combination of both.

- Depending on the analysis of his data, he will know if his hypothesis is supported or rejected.
- Analyzing this data is the most important part to support his hypothesis.

6. Conclusion

- A report will need to be made with findings of the research.
- The research can give the theories and literature that support his research.
- He can make suggestions or recommendations for further research on his topic.

Role of Theory in Behavioural Research:

1. Theory as orientation

- Narrows the range of facts to be studied.
- Defining what kind of facts are relevant for study.

2. Theory as conceptualization and classification

- It provides conceptual structures and relationships with other concepts leading to various systems of conceptualization and classification.

3. Task of theory is summarizing

- Theory concisely summarizing of what is already known about the object of study.

4. Theory predicts facts

- Role of theory has several facts . Most obvious is the extrapolation from known to the unknown.

5. Theory points to gap in our knowledge

- Summarizes the known facts and predicts facts which have not yet been observed.

Approaches for Identifying Concepts Constructs and Variables

Concept	Variable	Construct
Empirical abstraction to classify natural phenomenon.	When concepts can be assigned with values, they can be manipulated as Variables , so their relationship can be examined.	Some concepts are observable and easily distinguished from others, but other concepts are less tangible and can be defined only by inference. Construct are invited names for abstract variables that cannot be measured directly , but by measuring relevant correlated behaviour or observable.

Concept

Definition

- ❖ Generalized properties or characteristics associated with objects , events or people is a concept
- ❖ Concept refers to something that is conceived of in the mind.

- Webster dictionary

Abstractions of particular aspect of human behaviour and characteristics are called as concepts.

- Polit & Beck

Types of Concepts

1. Concrete concepts

- ❖ Symbolize material objects , which can be seen , touched, and felt.
- ❖ We can point out numerous objects and measure their dimensions.
- ❖ Ex : book , table , building

2. Abstract concepts

- ❖ Refer to properties or characteristics of objects,
- ❖ Concepts are abstract or general ideas , which are important to how we think about particular subjects or ideas.
- ❖ Ex: attitude , intelligence , leadership.

We can measure concepts through direct and indirect observations:

Direct Observation :

- ❖ We can measure someone's weight or height .
- ❖ we can record the colour of their hair or eyes.

Indirect Observation:

- ❖ we can use a questionnaire in which respondents provide answers to our questions about age, gender ,income , age , attitudes and behaviour.

Construct

Definition

- ❖ Higher order concepts are called as constructs.
- ❖ Constructs are abstractions that are deliberately invented / constructed by researchers.

- Kerlinger & Lee

- ❖ Like a concept a construct is an abstraction inferred from situations or behavior.

- Polit & Beck

- ❖ Constructs are measured with multiple variables.
- ❖ Constructs are considered latent variable because they cannot be directly observable or measured.
- ❖ Brand loyalty is a construct that marketing researchers study often ;

It can be measured using a variety of measures:

- ❖ Monetary value of past purchases.
- ❖ No of items purchased in the past.
- ❖ Frequency of past purchase occasions.
- ❖ The likelihood of switching to a competitive brand.

Differences Between Concept and Construct

CONCEPT	*CONSTRUCT*
Classifying and categorizing objects/events that have common characteristics beyond any single observation creates concepts.	It is a concept with scientific purpose or theoretical definition of a concept.
Concepts have progressive level of abstraction.	Construct is an image/ abstract idea specifically invented for a given research.
We can measure concepts directly and indirectly.	We can measure construct with multiple variables.
Example: Intelligence	Example : IQ

Role of Concept in Social Research

- ❖ They simplify thinking by pooling no of events under general heading
- ❖ To facilitate better communication by providing a common meaning.
- ❖ We design hypothesis using concepts. We devise measurement concepts by which to test these hypothetical statements.
- ❖ The success of research hinges on :
  - ▷ how clearly we conceptualize and
  - ▷ how well others understand the concepts we use.
- ❖ The challenge is to develop concepts that others will clearly understand.

Role of Construct in Social Research

- ❖ A Construct is an abstract idea inferred from specific instances that are to be related.
- ❖ Typical marketing constructs are brand loyalty ,satisfaction ,preference , awareness ,knowledge.
- ❖ Research objectives typically call for the measurement of constructs.
- ❖ There are customary methods for defining and measuring constructs.

References

Daivadeenam. L.L.Somani.Research methodology in extension education .

Patrick .Conceptual framework by Regoniel.

Sarah Vinz , 2015.Theoretical framework for your research.

C.R.kothari & Gaurav Garg 3r edition 2014.Research methodology methods and techniques.

E - Reference

https://www.questionpro.com/blog/empirical-research/

https://www.hamrolibrary.com/2021/05/concepts-in-social-research.html?m=1

https://www.hamrolibrary.com/2021/05/concepts-in-social-research.html?m=1

https://www.scribbr.com/dissertation/theoretical-framework/

https://www.researchgate.net/publication/322204158_THEORETICAL_AND_CONCEPTUAL_FRAMEWORK_MANDATORY_INGREDIENTS_OF_A_QUALITY_RESEARCH

Questions

1. What is Theoritical framework?
2. What is Conceptual framework?
3. What is Empirical framework?
4. How to take a literature review?
5. What is the purpose of review?
6. Define - Concepts.
7. What all are the types of Concepts?
8. Define - Construct.
9. What is the role of concept in social research?
10. What is the role of construct in social research?

Fill In The Blanks

1. ________is the structure or support of a theory of research study.
2. The key concepts of Theoretical frame work is _______.
3. Research should address the ______.
4. _______is the most important part to support the hypothesis.
5. _______ is the two ways to find the data analysis and result.
6. _______ is the empirical abstraction to classify natural phenomenon.

7. Concepts can be assigned with values, they can be manipulated as ______.
8. Concept refers to something that is conceived of in the mind, According to ______.
9. ______ is the example of concrete concepts.
10. We can measure someone's weight or height by ______

Observation

11. Higher order concepts are called as ______.
12. ______is a construct that marketing researchers study often.
13. Constructs are measured with ______.
14. ______&______are the examples of concept and construct.
15. The success of research hides on______.

Chapter 12

Variables

Introduction

- **Williams (1986)** defines a variable as "An observable characteristic of an object or event that can be described according to some well-defined classification or measurement scheme"
- A key element is that variables refer to characteristics that are not fixed but are able to vary, that is, to take on more than one value.
- For example, the word "**green"** would not be a variable but **"shades of green**" could be a variable. **"One inch"** is not a variable, however, **"length",** which could be operationally defined as the number of inches as measured by a ruler would be a variable.

Definition

- 'A variable, as the name implies, is" **something that varies".**
- It may be weight, height, anxiety levels, income, and body temperature and so on. Each of these properties varies from one person to another and also has different values along a continuum.
- A 'Variable' means some **characteristics of each number of the unit** that is to be studied such as **income, age, test score, land holding, etc.**
- Expressed numerically

Following are some examples of variables and some of their potential values.

Quantitative vs. categorical variables

- Data is a specific measurement of a variable – it is the value you record in your data sheet. Data is generally divided into two categories:
  - ▷ Quantitative data represents amounts.
  - ▷ Categorical data represents groupings.
- A variable that contains quantitative data is a quantitative variable
- A variable that contains categorical data is a categorical variable.

Quantitative variables

There are two types of quantitative variables: Discrete and Continuous

Type of variable	*What does the data represent?*	*Examples*
Discrete variables (aka integer variables)	Counts of individual items or values.	Number of students in a class Number of different tree species in a forest
Continuous variables (aka ratio variables)	Measurements of continuous or non-finite values.	Distance Volume Age

Categorical variables

- **Measured in categories.**
  - ▷ Categorical variables represent groupings of some kind.
  - ▷ For example, cars might be categorized as domestic or imported.
  - ▷ Categories must be mutually exclusive and exhaustive.
  - ▷ There are three types of categorical variables: **binary, nominal, and ordinal variables.**

Binary vs. nominal vs. ordinal variables

Type of variable	*What does the data represent?*	*Examples*
Binary variables (aka dichotomous variables)	Yes/no outcomes.	Heads/tails in a coin flip Win/lose in a football game
Nominal variables	Groups with no rank or order between them.	Species names Colors Brands
Ordinal variables	Groups that are ranked in a specific order.	Finishing place in a race Rating scale respon Rating scale responses in a survey*

Qualitative variables:

- This consists of categories **which can't be ordered in magnitude.**
- Qualitative variables comprise categories which do not have a quantitative relationship among themselves – Sex, race, religion, which can't be ordered in magnitude

Independent vs. dependent variables

- **Dependent variables**- it is one about which the experimenter makes a prediction.
- **Independent variables** – is one which is manipulated, measured and selected by the experimenter for the purpose of producing observable changes in the dependent variable

Type of variable	*Definition*	*Example (salt tolerance experiment)*
Independent variables (treatment variables)	Variables you manipulate in order to affect the outcome of an experiment.	The amount of salt added to each plant's water.
Dependent variables (response variables)	Variables that represent the outcome of the experiment.	Any measurement of plant health and growth: in this case, plant height and wilting.
Control variables	Variables that are held constant throughout the experiment.	The temperature and light in the room the plants are kept in, and the volume of water given to each plant.

Classification of Independent variables,

- **Task variables** – Those characteristics which are associated with a behavioral task presented to the subject
- **Environmental variables** – Those characteristics of the environment which are not the physical part of the task as such, but tend to produce changes in the behavioral measures. Temp, Noise, Illumination, time of the day are environmental variables
- **Subject variables** – Those characteristics of the subjects which are likely to produce changes in the behavioral measure. Sex, age, weight, anxiety, intelligence

Other variables,

- **Moderator variables** – which are manipulated and selected by the experimenter because they are suspected to moderate the relationship of the independent variables with the dependent variables
  - ▷ Study the relative effectiveness of lecture method and demonstration method upon class room achievement
  - ▷ Here two group of students, one with lecture and another with demonstration method
  - ▷ We can suspect that the intelligence of the students will also play a role here; we can now group the students in to low, and medium and high level of intelligence and each with lecture and demonstration method

- **Intervening variables –** those variables which theoretically exist and tend to influence the behavioral measure. They are hidden variables. They can't be seen and can't be manipulated by the experimenter – Anxiety/fatigue/ motivation
- **Active variable**– Variables which are manipulated are called as active variables- Rewards, punishments. Method of teaching
- **Attribute variables** – Can't be manipulated but can be measured – Age, sex, intelligence, race, are human characteristics which already been determined, can't be manipulated but can measure the performance of ten years old girl with 15 years old girl.

Reference

https://www.scribd.com/document/501948958/variables-types

Daivadeenam P.D, Somani L.L. Research methodology in Extension education. 2nd Edition: Agrotech publisher; 2004.

Nakkiran S, Ramesh G. Research methods in rural development. Deep and deep publication; 2010.

https://www.scribbr.com/methodology/types-of-variables/

Questions

1. What is the definition of a variable in research?
2. How many main types of variables exist in research, and what are they?
3. What is an independent variable?
4. What is a dependent variable?
5. Define control variables in research.
6. What is the first step in identifying variables in a study?
7. Why is operationalizing variables important in research?
8. How do you differentiate between discrete and continuous variables?
9. What is the role of extraneous variables in research?
10. Why is identifying variables crucial for research design and analysis?

Multiple Choice Questions

1. **What are variables in research?**

   a) Fixed constants

   b) Symbols used in equations

   c) Measurable factors that can vary

   d) Experimental errors

2. **How many main types of variables are commonly used in research?**

   a) Two

   b) Three

   c) Four

   d) Five

3. **Which of the following is an example of an independent variable?**

   a) Temperature in a chemistry experiment

   b) Student's age in an education study

   c) Income in a population survey

   d) Gender in a psychology study

4. **The variable that is observed and measured to assess the effect of the independent variable is called:**

   a) Extraneous variable

   b) Dependent variable

   c) Control variable

   d) Confounding variable

5. **Control variables are used to:**

   a) Manipulate the experimental conditions

   b) Ensure all participants are the same

   c) Keep other factors constant

   d) Measure the primary outcome

6. **Which is the first step in identifying variables in a research study?**

   a) Collect data

   b) Define the research question

   c) Conduct statistical analysis

   d) Review the literature

7. **Operationalizing variables means:**

   a) Assigning random values to variables

   b) Defining how variables will be measured or manipulated

   c) Creating control groups

   d) Eliminating variables from the study

8. **Discrete variables are characterized by:**

   a) An infinite number of possible values

   b) Only whole numbers as possible values

c) A continuous range of values

d) Non-numerical categories

9. **Extraneous variables are:**

a) Variables that are not relevant to the study

b) Variables that are controlled for in the research design

c) Variables that are always considered dependent

d) Variables that are never encountered in research

10. **Why is identifying variables crucial for research design and analysis?**

a) It makes the research more complicated

b) It ensures the research is less accurate

c) It helps to clarify the purpose and structure of the study

d) It has no impact on the research process

Chapter 13

Propositions and Hypotheses

What is proposition and hypothesis?

Proposition: A proposition is a statement or assertion that expresses a judgment or opinion about something. It can be either true or false.

A proposition is a theoretical statement that two or more factors or concepts are related and the type of relationship it is. It is a belief that may or may not have been tested.

Hypothesis: A hypothesis is an empirically testable version of a proposition. It is testable ,tentative explanation or prediction about a phenomenon.

It is a tentative statement showing the relationship between two or more variables.

Definition of hypothesis

- **James E Greighton-"** It's a tentative supposition or provisional guess which seem to explain the situation under observation"
- **Lundberg** defines "that hypothesis is a hunch, guess, imaginative idea which become the basis for action or investigation".
- **Webster** quotes that "Hypothesis is a proposition or condition or principle which is assumed perhaps without beliefs in order to draw out its logical consequences and by this method to test its accord with facts which are known or may be determined".
- **Black and champion-**"A tentative statement about something,the validity of which is usually unknown"
- **Kerlinger-**" A hypothesis is a conjectural statement of the relationship between two or more variables" A hypothesis is a hunch , assumption, suspiction or an idea about a phenomenon,relationship or situation,the reality or truth of which you don't know

Types of hypothesis

The following are the types of hypothesis,

- ❖ Null hypothesis
- ❖ Alternate hypothesis
- ❖ Directional hypothesis
- ❖ Non Directional hypothesis
- ❖ Simple hypothesis
- ❖ Complex hypothesis
- ❖ Research hypothesis
- ❖ Statistical hypothesis

Null Hypothesis (H0): The null hypothesis is a statement that there is no significant difference or relationship between variables. It is often used to test the absence of an effect.

Alternative Hypothesis (Ha or H1): The alternative hypothesis is the opposite of the null hypothesis. It suggests that there is a significant difference or relationship between variables..

Directional Hypothesis: This type of hypothesis specifies the direction of the expected relationship between variables. For example, "The more time a person spends exercising, the less body fat they will have."

Non-Directional Hypothesis: Also known as a two-tailed hypothesis, this type of hypothesis does not specify the direction of the expected relationship between variables. It only suggests that a relationship exists. For example, "There is a relationship between the amount of sleep and test performance."

Simple Hypothesis: A simple hypothesis posits a single relationship between two variables. For example, "Increased fertilizer will lead to larger plant growth."

Complex Hypothesis: A complex hypothesis suggests multiple relationships between variables. For example, "The interaction of temperature and humidity will affect plant growth differently in different types of soil."

Research Hypothesis: This type of hypothesis is specific to a particular research study. It is formulated based on the research question and the available evidence. It is what the researcher intends to investigate.

Statistical Hypothesis: In experimental research, statistical hypotheses involve statements about the population parameters. They often include terms like means, proportions, variances, and correlation coefficients.

Characteristics of good hypothesis

- ❖ **Testability-**A hypothesis should be testable,empirically tested
- ❖ **Clarity-** A hypothesis should not be ambiguous

- **Relevence-**A hypothesis must be self relevant in relation to the problem understudy and related to body of theory
- **Consistency-**Hypothesis must be self consistent without internal contradictions. It must be precise and specific
- **Simplicity-** Hypothesis must be simple without involving many complexities and assumptions
- **Adjustment-** A hypothesis should fit the facts and also the to the technique of data collection
- **Conformity to the law of parsimony-** The hypothesis should have the power of explaining the greatest possible number of observable facts without the necessity of its decision *i.e.*, it should confirm to thr law of parsimony
- **Fruitful for new discoveries-** A hypothesis should suggest new field of knowledge to be suited
- **Verification-**The hypothesis should be capable of verification **x. Operationalisable-** A hypothesis should be operationalization. This means that it can be expressed in terms that can be measured. If it cannot be measured ,it cant be tested and,hence,no conclusions can be drawn

Importance of hypothesis

- **Guiding Research Design**: Hypotheses provide a clear direction for the research study. They help researchers define the scope of the study, select appropriate research methods, and determine which variables to investigate. This guidance is essential for designing a research project that is both focused and relevant.
- **Testable Predictions**: Hypotheses are testable predictions or educated guesses about the relationships between variables. They help researchers formulate specific research questions and predictions, which can be empirically tested through data collection and analysis.
- **Organizing Thoughts**: Hypotheses help researchers organize their thoughts and ideas. They encourage researchers to think critically about the topic and to articulate their expectations and assumptions clearly. This process of hypothesis formulation often leads to a deeper understanding of the research problem.
- **Hypothesis Testing**: The process of testing hypotheses allows researchers to determine whether their predictions are supported or contradicted by empirical evidence. This testing is essential for drawing valid and reliable conclusions from social research.
- **Scientific Rigor**: Hypotheses are a fundamental component of the scientific method. They provide structure and rigor to the research process by ensuring that research questions are specific and testable, data collection is systematic, and data analysis is objective.

- **Theory Development and Refinement**: In social research, hypotheses are often derived from existing theories or conceptual frameworks. Testing hypotheses can lead to the refinement of existing theories or the development of new ones. This contributes to the growth of knowledge in the field.
- **Communication and Replicability**: Hypotheses make research findings more accessible and understandable to other researchers and the broader academic community. When hypotheses are clearly stated, it is easier for others to replicate the study, verify the results, or build upon the research.
- **Practical Applications**: Social research often aims to solve practical problems or inform policy decisions. Hypotheses provide a structured approach to understanding and addressing these issues, leading to evidence-based solutions and informed decision-making.
- **Minimizing Bias**: Hypotheses can help researchers minimize bias by making their expectations explicit. This can aid in the objective interpretation of research findings. Researchers are more likely to report results accurately, even if they contradict their initial expectations.
- **Resource Allocation**: Hypotheses can help researchers allocate limited resources effectively. By focusing on specific research questions, hypotheses enable researchers to allocate time, budget, and effort to areas of inquiry that are most likely to yield meaningful results.
- **Contributing to Knowledge**: Hypotheses are the building blocks of scientific knowledge. When hypotheses are tested and confirmed, they contribute to the body of knowledge in the social sciences. Over time, these findings can lead to a deeper understanding of human behavior and societal phenomena.

Reference

"Research methodology-methods and techniques"-

C.R.Kothari,Gaurav Garg

"Social research methods-Qualitative and Quantitative approaches"-W.Lawrence neuman

"Research Methodology-A step by step guide for beginners"Ranjit kumar

"Research Methodology"-P.K.Sharma

"Research Methodology in extension education"-

P.Daivadeenam,L.L.Somani

Multiple Choice Questions

1. ---------------is an empirically testable version of a proposition.

a) Variables

b) **Hypothesis**

c) Fixed constants

d) Proposition

2." It's a tentative supposition or provisional guess which seem to explain the situation under observation"-The following definition was given by--------------

a) James E Greighton

b) Lundberg

c) Webster

d) Kerlinger

3. The ----------- hypothesis is a statement that there is no significant difference or relationship between variables.

a) Directional

b) Non directional

c) Alternate

d) **Null**

4. What suggests that there is a significant difference or relationship between variables?

a) Directional

b) Non directional

c) **Alternate**

d) Null

5. --------------- hypothesis specifies the direction of the expected relationship between variables.

a) Null

b) Alternate

c) **Directional**

d) Non directional

6. --------------hypothesisis known as a two-tailed hypothesis

a) Null

b) Directional

c) **Non directional**

d) Alternate

7. **Which hypothesis posits a single relationship between two variables?**
   a) Complex
   b) **Simple**
   c) Alternate
   d) Null

8. **What hypothesis suggests multiple relationships between variables?**
   a) Directional
   b) Non directional
   c) **Complex**
   d) Alternate

9. **Which is not the characteristic of good hypothesis**
   a) Relavance
   b) Testability
   c) Simplicity
   d) **Adaptability**

10. **" A hypothesis is a conjectural statement of the relationship between two or more variables" .The following statement is given by**
    a) James E Greighton
    b) Lundberg
    c) Webster
    d) Kerlinger

QUESTIONS

1. Define proposition
2. Define hypothesis
3. Differentiate between hypothesis and proposition
4. What are the characteristic of good hypothesis
5. What are the importance of hypothesis in the social research
6. List out the types of hypothesis
7. Differentiate between simple and complex hypothesis
8. Differentiate null and alternate hypothesis
9. Differentiate directional and non directional hypothesis
10. 10.What is null hypothesis

Chapter 14

Measurement and Levels of Measurement

Measurement-Meaning

- According to Campbell, Measurement means assignment of numbers to objects or events according to rules. Measurement is assigning numbers to objects or events according to rules (Stevens, 1946).
- According to Thorndike, anything that exists at all, exists in some quantity is capable of being measured.
- It is easy to assign number in respect of properties of some objects, but it is relatively difficult in respect of others.
- For instance, measuring such things as social conformity, intelligence or marital adjustment is much less obvious and requires much closer attention than measuring physical weight, biological age or person's assets.
- The purpose is to have information in a form in which variables can be related to each other.
- Using observations, rating scales or any other device that allows us to obtain information in a quantitative form is Measurement.
- It can be said that measurement is the process of quantification of some phenomena. Measurement means comparing an unknown quantity with a known quantity which is taken as unity.
- In social sciences, we make use of a physical (not necessarily mechanical) component or stimulus. Social science research follows a standardized procedure or mechanism, as is followed in the physical sciences.
- When a scale is applied on a person, it gives us a number (or symbol) on his attitude, IQ, interest, emotional stability, motivation, and so on.
- It follows, then, that a measurement operation is always a standardized way of proceeding, which may or may not make use of mechanical devices or stimuli, but which always results in classification of the objects measured into some non-overlapping categories labelled by numerals, or simply by symbol.

Functions of Measurement:

1) Allows to Summarize Information:

The framing of proper measures also allows the information to be summarized and presented in a better way. This also allows researchers to use various graphs, tables, and, charts to represent he data properly. This also makes the research and the research findings more presentable and attractive to any potential user of the research report.

2) Provides Better Understanding of a Situation :

Measurement allows better understanding of a situation as compared to a scenario where there is no measurement at all. For example, various data about population is obtained only when it is measured, Until and unless the data is measured, it does not provide in-depth understanding of the situation.

3) Allows to Quantify Data and Statistical Sophistication:

The process of measurement also allows the researcher to quantify the abstract variables and research parameters. The degree of statistical treatment of the data depends upon he measurement scale adopted to quantify the data.

4) Important for Research Approach:

Selection of measurement techniques also determines the research approach and the way a researcher will tend to solve the research problems. Deciding the measures is thus an essential part of the research activity. The selection of proper measures goes a long way towards making the research a better planned end organised activity.

5) Provides Important Set of Tools:

The measurement procedures and instruments to be used provide invaluable information to the researcher which allows him to reach at a decision regarding the research problem. It also has a bearing on the policies and programs. However. the measures that are framed are only the means towards the objective of the researcher and not the ends. It helps the Researcher in reaching at critical decisions regarding a research objective.

Importance Of Measurement

(i) Measurement is a purely descriptive process.

(ii) Measurement implies that the attributes of persons or objects are possessed in varying degrees and the degree of variation can be measured and represented.

(iii) Measurement, in essence, is a numerical process.

It is important for

i. Public opinion

ii. Polling agencies

iii. Commercial organization

For example, the study of beliefs and attitudes of Indian people towards illness and health measures or family size would help in instituting a social and educational program to mobilize the people towards vaccination or adoption of family planning.

Difference Between Physical and Psychological Measurement

Physical Measurement	*Psychological Measurement*
Absolute measurement	Relative measurement
Reference point is zero	Reference point is group performance
Perfection is reached somehow	Perfection is not possible
It has fixed unit for measurement (CONSTANT)	There are no any fixed units (DERIVED& INDEFINITE)
It has fixed tools and instrument	No fixed measuring tool for a variable.
Perfectly objective and valid	Subjective in nature
The data are directly interpreted	Raw data are meaningless and change into standard scores.
Almost at ratio scale	Nominal, ordinal and interval variable
Great precision. Eg: Length, Amount of substance, Temperature	Less precision Eg: Behavior, Subject's intelligence, Cognition

Levels of Measurement

1. Nominal Scale

On a nominal scale, we are able to classify objects or events into non overlapping categories purely on the basis of some qualitative character of the studied variable. Nominal scales are specifically helpful in exploratory work where the objective is to disclose relationship rather than obtain certain measurements The common sources that are tapped for such classification are listed below.

- Theoretical considerations (as in atomic structure).
- Observed similarities and differences in appearance/ behavior of the objects studied (The Periodic Table of Elements was developed from the observation that certain elements tend to give similar chemical reactions).
- Practical considerations (the attitude of a group of voters is better understood and explained if they are classified by their age, caste, sex, occupation etc.)

However, Nominal scale is so elementary and crude that many scientists would not like to consider it as measurement at all.

- Lowest level of Measurement

- Numbers are used to name, identify or classify persons, objects
- All members of one group are equivalent (reflexive, transitive and symmetrical)
- Members of any two groups are never equivalent
- Admissible statistical operations - Frequency, percentage, proportion, mode and co efficient of contingencies
- Addition, subtraction, multiplication, division are not possible
- Example: A sample of students may be classified as Hindus, Muslim, or based on sex, political party affiliation

2. Ordinal Scale

- Ordinal level measurement allows objects or events to be ordered by degree on the basis of possession of some characteristic that can be abstracted and measured quantitatively.
- Ordinal level measurement allows objects or events to be ordered by degree on the basis of possession of some characteristic that can be abstracted and measured quantitatively.
- Ex: Typical relationships among the classes are that the members of one class have more of some property or characteristic than those in other classes. Such a relationship is often designated by the symbol (>), which means 'greater than'.
- Drawbacks: ordinal measures are never absolute quantities nor do they convey that the distance between the different rank values is equal
- Since ordinal measurement is not equal interval measurements
- They do not incorporate absolute zero point
- There is no way to ascertain whether a person has none of the characteristics being measured
- EXAMPLE: Socio economic status of people

3. Interval Scale

- This scale is able to specify the ratio of any two intervals. It is, therefore, to be regarded as more powerful measurement compared to the three others already discussed.
- The starting point zero of the scale is arbitrarily chosen.
- Third level includes all the characteristics of the nominal and ordinal scales
- Numerically equal distance on the scale indicates equal distance indicated equal distance in the properties of the objects being measured
- And so, the unit of measurement is constant and equal interval measurement
- Four objects A, B, C, D given score of 20,16,8,4.

- Difference between A-C = 20-8 = 12 is equal to B-D = 16-4 = 12, Also A-B = 4, B-C = 4.

4. Ratio Scale

- Ratio scale provides the most powerful measurement for it satisfies not only all the characteristics of an interval scale, but has also additional and vital characteristics - that it has an invariant or absolute zero.
- Common example of ratio scales are measures of weight, width, length, loudness etc.,
- So it is obvious that ratio scales are common in physical sciences than social sciences
- In the ratio scales zero point is true.
- Example: Length as feet or inches – ratio scales.

SCALE TYPE	***EXAMPLES***
NOMINAL SCALE	Do you like the quality of health care services provided by your health centre Yes-1 No-2
ORDINAL SCALE	If you are asked to rate the quality of health care services provided by the primary health care centre, how will you rank it? Excellent-1 Very Good -2 Good-3 Average-4 Poor-5
INTERVAL SCALE	In the past twelve months, how many times have you gone for check up to the primary health Centre? <5 times – 1 5 to 11 times - 3 >11 times – 4
RATIO SCALE	In the past twelve months, how much of money have you spent on health care? Amount of money spent......

References

Kothari C.R. 2004. Research Methodology Methods and Techniques. Second edition: New Age International Publishers

Singh G.B. 2011. Research Methodology (Advanced Techniques with Statistical Methods). Paradise Publishers

Daivadeenam P.D, Somani L.L. 2004. Research Methodology in Extension Education. Agrotech Publishers

https://conjointly.com/kb/measurement-in-research/

https://www.researchgate.net/publication/340829715_MEASUREMENT_IN_RESEARCH

Multiple Choice Questions

1. **In psychological measurement the reference point is ____________.**

   a. zero b. group performance c. definite d. none of the above

2. **Which is not a level of measurement**

   a. nominal b. ordinal c. interval d. cardinal

3. **__________ scales are helpful in exploratory work where the objective is to disclose relationship rather than obtain certain measurements.**

   a. ratio b. ordinal c. nominal d. interval

4. **__________ scale is able to specify the ratio of any two intervals.**

   a. ratio b. ordinal c. nominal d. interval

5. **In research a YES/NO scale is __________.**

   a. nominal b. ordinal c. interval d. cardinal

6. **________ scale provides the most powerful measurement.**

   a. ratio b. ordinal c. nominal d. interval

7. **The application of ________ scale shows a statement of 'greater than' or 'less than' or 'equal to' without indicating the extent of greater or less.**

   a. nominal b. ordinal. c. interval d. cardinal

One Mark

1. Define Measurement.
2. List out levels of measurement.
3. Differentiate physical and physiological measurement.
4. Drawbacks of ordinal scale.
5. Functions of measurement.
6. Brief note on ordinal scale.
7. Explain interval scale with example.
8. Explain nominal scale with example.
9. Explain the importance of measurement
10. List out the functions of measurement.

Chapter 15

Research Designs

Introduction

Research design is the skeleton for the research project. Decisions regarding what, where, when, how much, by what means concerning an inquiry or a research study constitute a research design. It is a process of deliberate anticipation directed towards bringing an expected situation under control. It supports the study and hold it's together. .It refers to the researcher overall plan for answering the research question or testing the research hypotheses.

Meaning of Research Design

- A research design is the arrangement of conditions for collection and analysis of data in a manner that aims to combine relevance to the research purpose with economy in procedure.
- Research design is a set of advance decisions that make up the master plan specifying the methods and procedures for collecting and analyzing the needed information.
- The Planned sequence of the entire process involved in conducting a research study. - Miller

Why is Research Design Important ?

- Good research design is the "first rule of good research."
- Knowledge of the needed research design allows advance planning so that the project may be conducted in less time and typically at a cost savings due to efficiencies gained in pre-planning

Purposes of Research Design

- It provides the scheme for answering research question.
- It maintains control to avoid bias that may affect the outcomes.
- It organizes the study in a certain way defending the advantages of doing while being aware and caution about potential disadvantages.
- A research design is the arrangement of the conditions for collection and analysis of data in a manner that aims to combine relevance to the research

purpose with economy in procedure

- Helps take decisions about how to complete the entire research:
  - Validly
  - Objectively
  - Accurately
  - Economically

Objectives of Research Design

- To gain background information and to develop hypotheses
- To measure the state of a variable of interest
- To test hypotheses that specify the relationships between two or more variables

Advantages

- Lead to more accurate results.
- Give optimum efficiency and reliability.
- Minimize the wastage of time as well as money.
- Instills confidence in the research.
- Provides satisfaction & success

Components

- Title of the study
- Statement of the problem
- Review of literature
- Area & Scope of Study
- Objectives of the study
- Formulation of hypothesis
- Definition of concepts
- Methodology
- Sampling design
- Constructing the schedule/Questionnaire
- Collection of data
- Analysis of data
- Interpretation of results
- Reporting the findings
- Time & Financial budgeting

Functions of Research Design

1. The identification and development of procedures & logistical arrangements required to undertake a study

- What procedures to follow and how to follow them
- Name the design e.g.'cross-sectional ' etc.,
- Provide detailed information about the following aspects of the study :
  - ▷ Who will constitute the study population ?
  - ▷ How will the study population be identified ?
  - ▷ Will a sample or the whole population be selected ?
  - ▷ If a sample is selected ,how will it be contacted ?
  - ▷ How will consent be sought?
  - ▷ What method of data collection will be used and why?
  - ▷ In the case of a questionnaire , where will the response be returned ?
  - ▷ How should respondents contact you if they have queries ?
  - ▷ In the case of interviews , where will they be conducted ?
  - ▷ How will ethical issues be taken care of?

2. Ensuring that the procedures are adequate in quality to obtain valid , objective & accurate answers to the research questions . Kerlinger calls it as "control of variance "

- To ensure that independent variable has the maximum opportunity to have its effect on the dependent variable
- Its important to select a study design that helps isolate , eliminate or quantify the effects of different sets of variable influencing the independent variable .

Classification of Study Designs

Three types of classification on the basis of :

- The number of contacts with the study population
- The reference period of the study
- The nature of the investigation

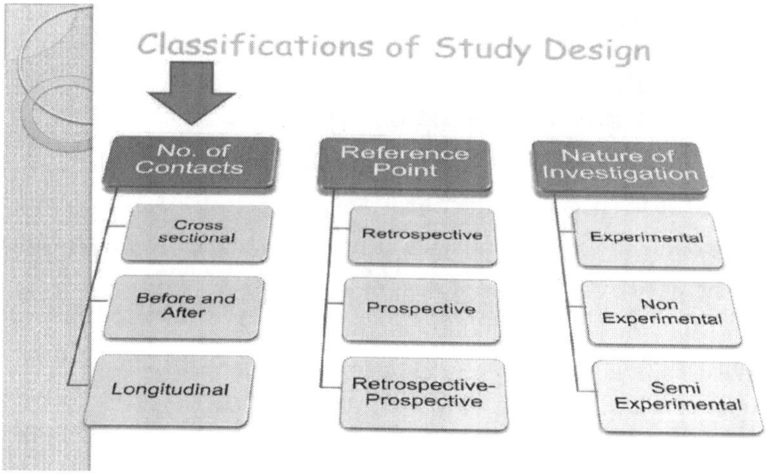

Study Design Based on Number of Contacts with the Population

1. Cross sectional studies

- It's a one shot or status studies / one time contact. Cross sectional studies are most commonly used design in social sciences to obtain overall picture.
- It's a simple design that decide what you want to find out about, identify the study population, select a sample & contact your respondents to find out the required information.
- Best for finding out about the phenomenon, situation, problem , attitude & issue .
- **Examples:** the health needs of a community , the attitude of students towards the facilities available in their library , consumer satisfaction with a product.
- **Advantages** : easy to analyse
- **Disadvantage**: this cannot measure the change.

2. Before & after study designs:

- Pre-test /post- test design
- Two points of contact with the same population
- Best suited to measure extent of change in situations , phenomenon , problems
- **Examples :** impact of increased funding on the quality of teaching in universities & affects of an advertisements on the sale of a product
- **Disadvantages :** it requires a longer time to complete , the lapse in time may cause , changes , maturation or withdrawal of the original population .

3. Longitudinal design :

- The longitudinal design helps to study the pattern of change in relation to time
- Useful when factual information on a continuing basis is needed

- It's a information gathering technique
- Reliable in accuracy as data is collected on regular basis

Study Design Based on Reference Period

- It refers to time frame in which a study is exploring a phenomenon.

1. Retrospective design :

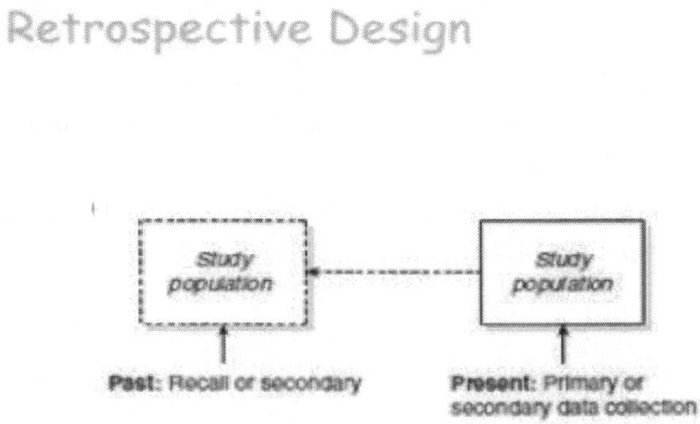

- Investigates a phenomenon, situation or problem that has happened in the past
- Examples :migratory movements in the subcontinent at the time of partition

2. The prospective study design :

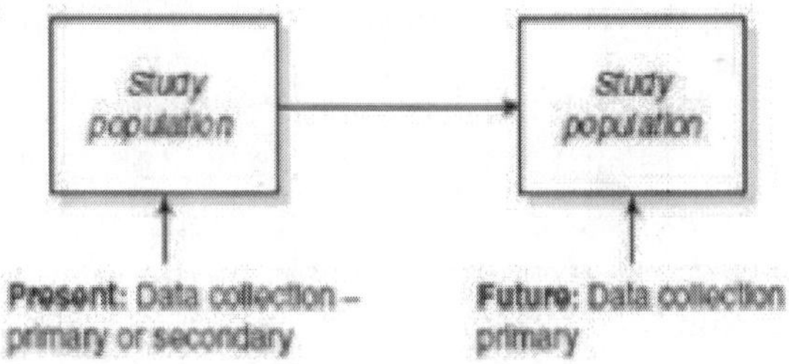

- This refers to the likely prevalence of a phenomenon, situation, problem or outcome in the future .
- Such studies attempt to establish the outcome of an event or what is likely to happen
- Experiments are usually classified as prospective studies as the researcher must wait for an intervention to register its effect on the study population.

3. Retrospective - prospective study design:

- It focuses on the past trends in a phenomenon and study them into the future
- Data collection is done in stages like retrospectively from existing records before the intervention is introduced
- The study population is followed to ascertain the impact of intervention.

Study Design Based on the Nature of the Investigation

Nature of investigation: whether the independent variable whose effect is to be investigated on a given phenomenon or situation is controlled or manipulated or not.

1. Experimental design:

- The researcher introduces the intervention that is assumed to be cause of a change and waits until it has produced or has been given sufficient time to produce the change .
- The independent variable can be observed , introduced , controlled or manipulated by the researcher
- An experimental study can be carried out in either a controlled or natural environment.

2. Non – experimental design:

- The researcher starts from the effects or outcomes and attempts to determine the causes.
- In this case the independent variable cannot be observed, introduced or controlled as the assumed cause has already occurred.
- Instead the researcher retrospectively links the cause to outcomes.

3.Semi-experimental design :

- It has the properties of both experimental and non-experimental studies

Research Design can be Split Into 4 Phases

- **Sampling design:** It deals with the method of selecting items to be observed for the given study
- **Observational design:** It relates to the conditions under which the observations are to be made
- **Statistical design:** It deals with the question of how many subjects are to be observed and how the observations are to be analyzed
- **Operational design:** It deals with the specific techniques by which the procedures specified in the sampling, statistical and observational designs can be carried out.

Different Research Designs

- Exploratory type research design
- Descriptive type research design
- Experimental type research design

Exploratory Type Research Design

- Explorative research studies are also termed as formulative research studies.

- Exploratory study is a systematic scientific approach which enables a social scientist to determine whether an idea is in reality or not.
- Explorative studies can provide ideas, hypothesis, suggestions that might never occur to the social scientist sitting in an office and mediating over the problem.& It is More flexible.
- It is Done in the field whether there is little knowledge is available.
- Exploratory studies which do not sets limits for themselves have limits imposed by various practical matters.
- Mostly the results obtained through the explorative study are to be treated as a sign post for future and further study in the same or similar direction. For this reason, they are also known as formulative studies.

Exploratory Research Methods:

- **Secondary data analysis**: the process of searching for interpreting existing information relevant to the research topic
- **Experience surveys**: refer to gathering information from those to be knowledgeable on the issues relevant to the research problem
- **Key-informant technique**: gathering information from those thought to be knowledgeable on the issues relevant to the problem
- **Lead-user survey**: used to acquire information from lead users of a new technology
- **Case analysis**: a review of available information about a former situation(s) that has some similarities to the current research problem
- **Focus groups**: small groups brought together and guided by a moderator through an unstructured, spontaneous discussion for the purpose of gaining information relevant to the research problem

Uses of Exploratory Research :

- Gain background information
- Define terms
- Clarify problems and hypothesis
- Establish research priorities

Descriptive type Research Design

- Rigid Design.
- Probability sampling design.
- Pre-planned design for analysis.
- Structured instruments for collection of data.
- Advanced decisions about operational procedure.
- Research design in which the major emphasis is on determining the frequency with which something occurs or the extent to which two variables co vary.

- Enable researcher to describe picture of a phenomenon under investigation.
- Methodology involved – qualitative in nature producing descriptive data.
- Three approaches to enable to record/analyse the behavioral patterns:
  - Participant observation.
  - Personal documents.
  - Unstructured interviewing.

Experimental Type Research Design

An experiment is defined as manipulating an independent variable to see how it affects a dependent variable while also controlling the effects of additional extraneous variables.

Experimental Type Research Design

- Professor R.A.Fisher's name is associated with experimental designs.
- It's origin in agricultural research was made by him when he was working in Rothamsted Experimental Station (Centre for Agricultural Research in England).
- Experiment is a study in which the investigator manipulates or varies (called the independent variables) & measures other variables (called the dependent variables).
- When an experiment is possible it is the most effective method of testing a hypothesis. i.e; one variable 'X' casually influences another variable 'Y'
- A blueprint of the procedure that enables the researcher to maintain control over all factors that may affect the result of an experiment.
- The researcher attempts to determine or predict what may occur.
- Experimental research is often used where there is time priority and the magnitude of the correlation is great.
- The classic experimental design specifies an experimental group and a control group.
- Subsequent experimental designs have used more groups and more measurements over longer periods. True experiments must have control, randomization, and manipulation.

There are three basic principles of experimental designs :

1. The principle of replication
2. The principle of randomization
3. The principle of local control

1. The Principle of Replication

According to this , the experiment should be repeated more than once. Thus, each treatment is applied in many experimental units instead of one. By doing so, the statistical accuracy of the experiments is increased.

2. The Principle of Randomisation

This provides protection against the effects of extraneous factors by randomisation. We may apply randomisation principle and protect ourselves against the effects of the extraneous factors.

3. The Principle of Local Control

According to this principle, we first divide the field into several homogeneous parts, known as blocks, and then each such block is divided into part equal to the number of treatments.

Features of a Good Research Design

A research design appropriate for a particular research problem, usually involves the following features.

- The mean of obtaining information.
- The availability and skills of the researcher and his staff, if any.
- The objective of the problem to be studied.
- The nature of the problem to be studied
- The availability of time and money for the research work.

Designs Associated with Quantitative Research

- The cross –over comparative experimental design
- Trend studies
- Cohort studies
- Panel studies
- Blind studies
- Double –blind studies

Design used in Qualitative Research

- Case study
- Oral history
- Focus group
- Participant observation
- Holistic research
- Community discussion forum
- Reflective journal log

Process or Methods of Diagnostic Research Studies

- Formulating the objective of the study.
- Designing the methods of data collection.
- Selecting the sample.

- Collecting the data.
- Processing
- Analysing the data.
- Reporting the findings.

Pretest and Post test

- **Pretest** refers to the measurement of the dependent variable taken prior to changing the independent variable.
- **Post test** refers to measuring the dependent variable after changing the independent variable.

Research Design vs Variance Control

Control of variance

The control variates method is a variance reduction technique used in Monte Carlo methods. It exploits information about the errors in estimates of known quantities to reduce the error of an estimate of an unknown quantity.

One of the most important function of experimental design is to control variance. The problem of variance control has three aspects, 1.Systematic Variance, 2. Extraneous Variance and, 3.Error Variance.

- The main functions of the experimental design, are to maximize, the effect of, systematic variance, control, extraneous, source of variance, and minimize, error variance.

Maxmincon Principle

One of the basic principles of research design is that of **maximizing the systematic or desirable variance, minimizing the error variance and controlling the effect of the extraneous variables**. This principle was referred to MAXMINCON by Kelinger and Pedhazur (1973) and Kerlinger (1986).

Criteria For Selecting A Suitable Research Design

Following are the factors to be considered while deciding your research methodology:

- Research Goal-Think of your research goals.
- Statistical significance- Another essential factor to consider while choosing the research methodology is statistical results.
- Quantitative vs qualitative data.
- Sample size.
- Timing.

Multiple Choice Questions

1. **What is the primary purpose of a research design?**

   a) Collecting data

   b) Testing hypotheses

   c) Structuring the research process

   d) Writing research reports

2. **Which type of research design is best suited for exploring new and unfamiliar topics?**

   a) Descriptive

   b) Experimental

   c) Exploratory

   d) Causal-comparative

3. **In experimental research, the variable that the researcher manipulates is called:**

   a) Dependent variable

   b) Extraneous variable

   c) Independent variable

   D Control variable

4. **Which of the following is a characteristic of a longitudinal research design?**

   a) Data collected at a single point in time

   b)Focus on cause-and-effect relationships

   c) Follows subjects over an extended period

   d) Emphasizes cross-sectional comparisons

5. **External validity in research design refers to the:**

   a) Consistency of measurement

   b) Generalizability of findings

   c) Precision of experimental control

   d) Stability of results over time

6. **What is the purpose of a pilot study in research design?**

   a) To test the statistical significance

   b) To estimate the sample size

   c) To refine and improve the research design

   d) To select the research variables

7. **Which of the following is a potential ethical concern in research design?**
   a) Data analysis techniques
   b) Sampling techniques
   c) Participant confidentiality
   d) Research hypotheses

8. **Triangulation in research design involves:**
   a) Using multiple methods to study the same phenomenon
   b) Conducting research in three different countries
   c) Reducing bias in data collection
   d) Combining qualitative and quantitative data

9. **What is the initial step in developing a research design?**
   a) Data collection
   b) Literature review
   c) Hypothesis testing
   d) Sampling

10. **Which of the following is a characteristic of a cross-sectional research design?**
    a) Data collected over an extended period
    b) Focus on cause-and-effect relationships
    c) Data collected at a single point in time
    d) Experimental manipulation

Questions

1. What is the purpose of Research Design?
2. Differentiate between exploratory, descriptive and explanatory research designs
3. What are the main advantages and disadvantages of qualitative research design?
4. Explain the concept of triangulation in research design.
5. What is the difference between cross- sectional and longitudinal research designs ?

Chapter 16

Quantitative Research Designs

Research Design

A research design is the plan or framework used to conduct a research study. It involves outlining the overall approach and methods that will be used to collect and analyze data in order to answer research questions or test hypotheses.

Important Concepts of Research Design

Variable

Variable is a concept that can take on various quantitative values. For instance, weight, height, etc.

Dependent Variable

A dependent variable is a variable that is tested in an experiment. It is dependent, in some way on the variation of an independent variable.

Independent Variable

An independent variable in an experiment is considered to stand on its own. For instance, if the test scores of a class are an outcome of their efforts; efforts are an independent variable, and the score is a dependent variable.

Hypothesis

It is defined as the proposition that needs to be tested in an experiment.

Research design is broadly divided into quantitative and qualitative research design:

Quantitative Research Design

Quantitative research design aims at finding answers to who, what, where, how, and when through the course of research. Moreover, the outcome of the quantitative analysis is easy to represent in the form of statistics, graphs, charts, and numbers.

Qualitative Research Design

Qualitative research design focuses on finding answers to how and why. It uses open-ended questions and helps the subjects express their views clearly.

Qualitative research is ideal for businesses that aim to understand customers' behaviour and requirements.

Uses of quantitative research design:

- Determine the effects of an intervention.
- It can isolate and evaluate the effects of an intervention the independent variable on a specific outcome and the dependent variable.
- Measures relationships between or among variables.
- Measures change , or lack of change , over time.

Experimental design

Experimental research design is a framework of protocols and procedures created to conduct experimental research with a scientific approach using two sets of variables. Herein, the first set of variables acts as a constant, used to measure the differences of the second set. The best example of experimental research methods is quantitative research.

Experimental research helps a researcher gather the necessary data for making better research decisions and determining the facts of a research study.

Example:

- The effect of positive reinforcement on attitude towards school.
- The effect of a systematic preparation and support system on children who were scheduled for surgery on the amount of psychological upset and cooperation.
- Descriptive research:

Descriptive research :

Descriptive research is a research method used to try and determine the characteristics of a population or particular phenomenon.

Using descriptive research you can identify patterns in the characteristics of a group to essentially establish everything you need to understand apart from why something has happened. Market researchers use descriptive research for a range of commercial purposes to guide key decisions.

For example you could use descriptive research to understand fashion trends in a given city when planning your clothing collection for the year. Using descriptive research you can conduct in depth analysis on the demographic makeup of your target area and use the data analysis to establish buying patterns.

A description of how parents feel about twelve-month school year.

A description of the extent to which elementary teachers use maths manipulatives. As a survey method, descriptive research designs will help researchers identify characteristics in their target market or particular population.

Correlational design

A correlational research design investigates relationships between variables without the researcher controlling or manipulating any of them.

A correlation reflects the strength and/or direction of the relationship between two (or more) variables. The direction of a correlation can be either positive or negative.

Positive correlation	Both variables change in the same direction	As height increases, weight also increases
Negative correlation	The variables change in opposite directions	Coffee consumption increases while tiredness decreases
Zero correlation	There is no relationship between the variables	Coffee consumption is not correlated with height

Comparative research design

Comparative research essentially compares two groups in an attempt to draw conclusions about them. Researchers attempt to identify and analyze the similarities and differences between groups , and these studies are most often comparing two separate people groups. The researcher's goal is to determine whether the independent variable affected the outcome , or dependent variable by comparing two or more group of individuals. In this type of research investigators attempt to determine the cause or consequences of differences that already exist between or among group of individuals. It attempts to identify cause and effect relationships. Involve two or more group variables.

Example:

Some of the common variables investigated in this type of research are:

a) Achievement and other ability variables

b) Family-related variables

c) Variables related to school

d) Personality variables

Survey Research Design

Survey research refers to a particular type of research design where the primary method of data collection is by survey. In this study design, surveys are used as a tool by researchers to gain a greater understanding about individual or group perspectives relative to a particular concept or topic of interest. A survey typically consists of a set of structured questions where each question is designed to obtain a specific piece of information.

Survey research can be undertaken for a variety of reasons, but a common theme with surveys is that they are an easily accessible way for respondents to share or demonstrate their knowledge or perspectives about a particular topic.

Longitudinal survey research

Longitudinal survey research involves conducting survey research over a continuum of time and spread across years and decades.

Respondent behavior, preferences, and attitudes are continuously observed over time to analyze reasons for a change in behavior or preferences. For example, suppose a researcher intends to learn about the eating habits of teenagers. In that case, he/she will follow a sample of teenagers over a considerable period to ensure that the collected information is reliable

Cross-sectional survey research:

Researchers conduct a cross-sectional survey to collect insights from a target audience at a particular time interval. This survey research method is implemented in various sectors such as retail, education, healthcare, SME businesses, etc. Cross-sectional studies can either be descriptive or analytical. It is quick and helps researchers collect information in a brief period.

Ex Post Facto Research Design

An ex post facto research design is a method in which groups with qualities that already exist are compared on some dependent variable. Also known as "after the fact" research, an ex post facto design is considered quasi-experimental because the subjects are not randomly assigned - they are grouped based on a particular characteristic or trait.

For example, a researcher is interested in how weight influences self-esteem levels in adults. So the participants would be separated into differing groups (underweight, normal weight, overweight) and their self esteem levels measured. This is an ex post facto design because a pre-existing characteristic (weight) was used to form the groups.

Limitations:

- Independent variables cannot be manipulated.
- Risk of improper interpretation of results.
- Lack of scientific study regarding cause and effect relationship.
- Cannot randomly assign the subjects to different groups.

Secondary Data Analysis

Secondary data analysis involves a researcher using the information that someone else gathered for his or her own purposes.

Researcher leverage secondary data analysis in an attempt to answer a new research question , or to examine an alternative perspective on the original question of a previous study.

Steps in Secondary Data Analysis:

- **Determine your research question** – As indicated above, knowing exactly what you are looking for
- **Locating data**– Knowing what is out there and whether you can gain access to it. A quick Internet search, possibly with the help of a librarian, will reveal a wealth of options.
- **Evaluating relevance of the data** – Considering things like the data's original purpose, when it was collected, population, sampling strategy/sample, data collection protocols, operationalization of concepts, questions asked, and form/shape of the data.
- **Assessing credibility of the data** – Establishing the credentials of the original researchers, searching for full explication of methods including any problems encountered, determining how consistent the data is with data from other sources, and discovering whether the data has been used in any credible published research.
- **5.Analysis** – This will generally involve a analyzing the data of range with statistical methods applied.

Questions

1. Define Research Design.
2. State the types of correlations with appropriate examples.
3. Explain Survey Research design along with their types .
4. How does descriptive research varies from experimental research?
5. Why Ex Post Facto is called "after the fact" research – Justify
6. Explain Comparative Research Design.
7. Explain correlational Research Design.
8. What are the uses of Quantitative Research Design?
9. What are the steps in secondary data analysis?

Objectives

1. There is no relationship between the variables is _______ correlation
2. __________ research designs will help researchers identify characteristics in their target market or particular population in research study
3. Respondent behavior, preferences, and attitudes are continuously observed over time to analyze reasons for a change in behavior are examples for _________ survey design.
4. An ________ research design is a method in which groups with qualities that already exist are taken for the study.

5. Both variables change in the same direction in _______ correlation
6. Evaluating relevance of the data is one of the step in __________
7. _________ is the plan or framework used to conduct a research
8. _________ survey is used to collect insights from a target audience at a particular time interval.
9. A_______research design investigates relationships between variables without the researcher controlling or manipulating any of them.
10. __________ is defined as the proposition that needs to be tested in an experiment.

Chapter 17

Qualitative and Mixed Methods Research Designs

Qualitative Research Design

This type of research is quite contrary to quantitative research design. It is explanatory in nature and always seeks answers to "What's" and "How's".

- ❖ It mainly focuses on why a specific theory exists and what would be the respondent's answer to it. This allows a researcher to draw a conclusion with proper findings.
- ❖ Although the processes are similar, qualitative methods rely on text and image data, have unique steps in data analysis, and draw on diverse designs.
- ❖ Due to this, qualitative research is often defined as being subjective.

Qualitative Research Designs

- ❖ Phenomenological Studies
- ❖ Ethnographic Studies
- ❖ Grounded Theory Studies
- ❖ Historical Studies
- ❖ Case Studies
- ❖ Action Research Studies

Ethnographic

Ethnographic studies involve the collection and analysis of data about cultural groups. Agar (1986) described ethnography as "encountering alien worlds and making sense of them" . He further stated that ethnographers try to show how actions in one world make sense from the point of view of another world.

- ❖ In ethnographic research, the researcher frequently lives with the people and becomes a part of their culture.
- ❖ The researcher explores with the people their ***rituals and customs***. An entire cultural group may be studied or a subgroup in the culture.
- ❖ The term culture may be used in the broad sense to mean an entire tribe of Indians, for example, or in a more narrow sense to mean one nursing care unit.

- Ethnographers interview people who are most knowledgeable about the culture. These people are called key informants.
- A classic example of ethnographic research would be **an anthropologist traveling to an island, living within the society on said island for years, and researching its people and culture through a process of sustained observation and participation.**
- Gance-Cleveland (2004) examined the features, critical attributes, processes, and benefits of school based support groups for adolescents with an addicted parent. Ethnographic methods were used to gather data. Participant observations were conducted weekly at two high schools over one semester. Interviews were conducted with program administrators, school administrators, group co-facilitators, and participants. School-based support group participation was found to enhance self-knowledge and led to self-care and self-healing

Grounded Theory

Grounded theory is a qualitative research approach developed by two sociologists, Glaser and Strauss (1967). Grounded theory studies are studies in which data are collected and analyzed and then a theory is developed that is grounded in the data. Some of the terms used by Glaser and Strauss are difficult for nurses to understand. Leininger (1985) wrote that in 1980 she began to translate their terms into what she called "standard English.

- Grounded theory is a qualitative research approach developed by two sociologists, Glaser and Strauss (1967). Grounded theory studies are studies in which ***data are collected and analyzed*** and then a ***theory is developed*** that is grounded in the data.
- Examples: Grounded theory is often used by the HR department. For instance, they might **study why employees are frustrated by their work**. Employees can explain what they feel is lacking. HR then gathers this data, examines the results to discover the root cause of their problems and presents solutions.
- The grounded theory qualitative method was used by Williams and Irurita (2005) to study the personal control and emotional comfort of hospitalized patients. Interviews were conducted with 40 patients, and 75 hours of field observations were conducted. The basic psychological process identified by the researchers was labeled "optimizing personal control to facilitate emotional comfort." Personal control referred to the ability of patients to influence their environment; emotional comfort was defined as a state of relaxation that affected the physical status of the patient. Personal control was found to be a central feature of emotional comfort.

Phenomenological

- Phenomenological studies examine human experiences through the descriptions provided by the people involved. These experiences are called lived experiences.

- The goal of phenomenological studies is to describe the meaning that experiences hold for each subject.
- This type of research is used to study areas in which there is little knowledge (Donalek, 2004).
- In phenomenological research, respondents are asked to describe their experiences as they perceive them.
- Examples: Examples of phenomenological research include **exploring the lived experiences of farmers after a natural disaster leading to a heavy crop loss.** The term phenomenology often is used without a clear understanding of its meaning.
- Phenomenological Study :Daly (2005) studied the lived experiences of mothers of suicidal adolescents. She contended that, unfortunately, the mother's experience is often the hidden dimension in the family. Unstructured interviews were conducted with 6 mothers living with suicidal adolescents. Six themes were identified: failure as a good mother, the ultimate rejection, feeling alone in the struggle, helplessness and powerlessness in the struggle, cautious parenting, and keeping an emotional distance.

Narrative Research

- A literary form of qualitative research, narrative research is all about collecting and telling a story or stories (in detail).
- Researchers write narratives about experiences of individuals, describe a life experience, and discuss the meaning of the experience with the individual.
- Usually, a narrative research design is focused on studying an individual person. The researcher becomes the interpreter of the individual's stories, as opposed to a community.
- Examples of narrative inquiry in qualitative research include for instance: **stories, interviews, life histories, journals, photographs and other artifacts**.

Mixed Method Designs

- Mixed methods research has the capacity to intersect with other approaches, such as action research, by adding a solid methodological foundation and creating an integrated approach for addressing complex practical problems.
- Combining the two approaches can produce more scientifically sound and transferable results by synergistically integrating qualitative stakeholder engagement with quantitative outcomes to inform action/intervention planning, implementation, evaluation, and monitoring.
- Mixed methods and action research are aimed at providing comprehensive information: **mixed methods seeks to provide comprehensive answers to study research questions, whereas action research seeks to provide comprehensive solutions to practical problems**.

Action Research Designs

Action Research Studies Action research is a type of qualitative research that seeks action to improve practice and study the effects of the action that was taken (Streubert & Carpenter, 2002). Solutions are sought to practice problems in one particular hospital or health care setting. There is no goal of trying to generalize the findings of the study, as is the case in quantitative research studies.

In action research, the implementation of solutions occurs as an actual part of the research process. There is no delay in implementation of the solutions. Action research became popular in the 1940s. Kurt Lewin (1946) was influential in spreading action research. He came interested in helping social workers improve their practice. Although many of you may have heard of Lewin and his contribution to change theory, his involvement in action research is not as well known

- Action research can be defined as ***"an approach in which the action researcher and a client collaborate in the diagnosis of the problem and in the development of a solution based on the diagnosis".***
- In other words, one of the main characteristic traits of action research relates to collaboration between researcher and member of organisation in order to solve organizational problems.
- Action Research Action research was used with staff in one hospice and one nursing home setting in London (Dunckley, Aspinal, Addington-Hall, Hughes, & Higginson, 2005).The purpose of the study was to identify facilitators and barriers to the use of the Palliative Care Outcome Scale (POS). Staff took part in semistructured interviews,completed diaries,and participated in monthly meetings to give their opinions of what they thought were the facilitators and barriers to the implementation of the POS.
- Participatory action research (PAR) is a special kind of community-based action research in which there is collaboration between the study participants and the researcher in all steps of the study: determining the problem, the research methods to use, the analysis of data, and how the study results will be used. The participants and the researcher are co-researchers throughout the entire research study. According to Kelly (2005), PAR provides an opportunity for involving a community "in the development and assessment of a health program

Advantages of Action Research

- High level of practical relevance of the research;
- Can be used with quantitative, as well as, qualitative data;
- **Possibility to gain in-depth knowledge about the problem.**

Disadvantages of Action Research

- Difficulties in distinguishing between action and research and ensure the application of both;

- Delays in completion of action research due to a wide range of reasons are not rare occurrences
- Lack of repeatability and rigour.

Translational Research

- Translational research is a bidirectional process that involves multidisciplinary integration among basic, clinical, practice, population, and policy-based research. The goal of translational research is to speed up scientific discovery into patient and community benefit.
- The term has been used most commonly in life-sciences and biotechnology but applies across the spectrum of science and humanities. In the context of biomedicine, translational research is also known as **bench to bedside**. In the field of education, it is defined as research which translates **concepts to classroom practice.**

References:

Creswell, John W. Research Design: Qualitative, Quantitative, and Mixed Methods Approaches.

Choudhary, Dilip Kumar. Introduction to qualitative research methods.

https://research-methodology.net/research-methods/action-research/

https://www.statisticssolutions.com/types-of-qualitative-research-designs/

https://www.umsl.edu/~lindquists/qualdsgn.html

https://leverageedu.com/blog/research-design/

http://edrm600narrativedesign.weebly.com/#:~:text=A%20literary%20form%20of%20qualitative,the%20experience%20with%20the%20individual.

Multiple Choice Questions

1. **Which of the following types of studies is considered a qualitative study?**

   a) Correlational

   b) Ethnographic

   c) Comparative

   d) Methodological

2. **Grounded theory research was developed by two**

   a) nurses.

   b) physicians.

   c) psychologists.

   d) sociologists.

3. **Margaret Mead was a researcher from what discipline?**

   a) Anthropology

   b) Psychology

   c) Nursing

   d) Sociology

4. **Which of the following statements is true concerning the examination of historical research?**

   a) Internal criticism should be considered before external criticism.

   b) External criticism should be considered before internal criticism.

   c) Both internal and external criticism should be considered simultaneously.

5. **Case studies may concern an in-depth examination of**

   a) individuals.

   b) groups of peoples.

   C) institutions.

   d) all of the above.

6. **Consider this title: "The Lived Experience of Surviving a Tornado." This title would indicate which of the following types of qualitative research?**

   a) Ethnographic

   b) Phenomenological

   c) Historical

   d) Grounded theory

7. **The review of the literature in qualitative research is**

   a) never conducted prior to the beginning of the study.

   b) sometimes conducted prior to the beginning of the study.

   c) always conducted prior to the beginning of the study.

8. **Which type of qualitative research approach would be most appropriate to study nurses' involvement in bringing about a change in the type of forms used to record patient data?**

   a) Action

   b) Ethnographic

   c) Phenomenological

   d) Historical

9. **Teresa Christy, a nurse, was involved in what type of qualitative research?**

   a) Action

   b) Ethnographic

c) Historical

d) Phenomenological

10. Which type or types of qualitative research has or have been conducted by nurses?

a) Action

b) Ethnographic

c) Phenomenological

d) Historical

e) All of the above

Short answers

1) What are qualitative research designs ?

Qualitative Research Designs

▷ Phenomenological Studies

▷ Ethnographic Studies

▷ Grounded Theory Studies

▷ Historical Studies

▷ Case Studies

▷ Action Research Studies

2) What is phenomenological study ?

❖ Phenomenological studies examine human experiences through the descriptions provided by the people involved. These experiences are called lived experiences.

❖ The goal of phenomenological studies is to describe the meaning that experiences hold for each subject.

❖ This type of research is used to study areas in which there is little knowledge (Donalek, 2004).

❖ In phenomenological research, respondents are asked to describe their experiences as they perceive them.

3) Write about ethnographical study?

Ethnographic studies involve the collection and analysis of data about cultural groups. Agar (1986) described ethnography as "encountering alien worlds and making sense of them" . He further stated that ethnographers try to show how actions in one world make sense from the point of view of another world.

❖ In ethnographic research, the researcher frequently lives with the people and becomes a part of their culture.

❖ The researcher explores with the people their ***rituals and customs***. An entire cultural group may be studied or a subgroup in the culture.

- The term culture may be used in the broad sense to mean an entire tribe of Indians, for example, or in a more narrow sense to mean one nursing care unit.
- Ethnographers interview people who are most knowledgeable about the culture. These people are called key informants.

4) Write about narrative study ?

- A literary form of qualitative research, narrative research is all about collecting and telling a story or stories (in detail).
- Researchers write narratives about experiences of individuals, describe a life experience, and discuss the meaning of the experience with the individual.
- Usually, a narrative research design is focused on studying an individual person. The researcher becomes the interpreter of the individual's stories, as opposed to a community.
- Examples of narrative inquiry in qualitative research include for instance: **stories, interviews, life histories, journals, photographs and other artifacts**.

5) Define case studies .

- Case studies are in-depth examinations of people or groups of people.
- A case study could also examine an institution, such as hospice care for the dying.
- Case studies are in-depth examinations of people, groups of people, or institutions. Content analysis is the term used to indicate the examination of communication messages obtained in case studies, as well as in other types of qualitative studies.

6) What is historical studies ?

- Historical studies concern the identification, location, evaluation, and synthesis of data from the past.
- Historical data should be subjected to both external and internal criticism.
- External criticism is concerned with the authenticity of the data, whereas internal criticism is concerned with the accuracy of the data.

7) Write about grounded theory ?

In grounded theory studies, data are collected and analyzed, and then a theory is developed that is grounded in the data. A process called constant comparison is used, in which data are constantly compared to data that have already been gathered. Pertinent concepts are identified and assigned codes.

8) What is action research study ?

Action research is a type of qualitative research that seeks action to improve practice and study the effects of the action that was taken. Participatory action research (PAR) is a special kind of action research in which there is collaboration between the study participants and the researcher in all steps of the study.

9) Advantages and disadvantages of action research ?

Advantages of Action Research

- ❖ High level of practical relevance of the research;
- ❖ Can be used with quantitative, as well as, qualitative data;
- ❖ **Possibility to gain in-depth knowledge about the problem.**

Disadvantages of Action Research

- ❖ Difficulties in distinguishing between action and research and ensure the application of both;
- ❖ Delays in completion of action research due to a wide range of reasons are not rare occurrences
- ❖ Lack of repeatability and rigour.

10) What is translational research ?

- ❖ Translational research is a bidirectional process that involves multidisciplinary integration among basic, clinical, practice, population, and policy-based research. The goal of translational research is to speed up scientific discovery into patient and community benefit.

The term has been used most commonly in life-sciences and biotechnology but applies across the spectrum of science and humanities. In the context of biomedicine, translational research is also known as **bench to bedside**. In the field of education, it is defined as research which translates **concepts to classroom practice.**

Chapter 18

Elements of Research Design and Triangulation

Elements of Research Design

Research Design is a blueprint for data collection, measurement and analysis. It is the stepping stone for decisions regarding what, where, how much, when and steps for research design are undertaken.

The Key Elements That Every Good Research Design Have Are:

- Purpose statement.
- Data collection techniques.
- Methods of research data analysis.
- Type of research methodologies.
- Settings for research study.
- Time of the research study.
- Analysis measurement.
- Possible obstacles to the research.

Research Strategies

A research strategy refers to the step-by-step plan of action that gives direction to the researcher's thought process. It enables a researcher to conduct the research systematically and on schedule.

The main purpose is to introduce the principal components of the study such as research topic, areas, major focus, research design and finally the research methods.

An appropriate strategy has to be selected on the basis of the following:

- Research questions.
- Research objectives.
- Amount of time available.
- Resources at researcher's disposal.
- Philosophical underpinnings of the researcher.

Types of Research Strategies

1. QUALITATIVE:

- This strategy is generally used when to understand the **underlying reasons** or opinion of the people on certain **facts** or problem.
- It does not involve numerical data.
- It provides insides into the research problem and hence helps in achieving the research problem.
- Various methods that can be used include interviews, observations, open-ended surveys and focus group discussions.

Example- Unstructured interviews which generate qualitative data through the use of open questions. This allows the respondent to talk in some depth, choosing their own words. This helps the researcher develop a real sense of a person understanding of a situation.

2. Quantitative

- It involves the collection of **primary or secondary data** which is in numerical form.
- Under this strategy, the researcher can collect the data by using questionnaire, polls and the surveys mainly focuses on the when, what and how often a specific phenomenon occurs.

Example - Survey conducted to understand the amount of time a doctor takes to tend to a patient when the patient walks into the hospital.

3. Descriptive

- This is generally used when the researcher wants **to describe** a particular situation.
- This involves observing and describing the behaviour patterns of either an individual, community or any group.
- One thing that distinguishes it from other forms of research strategies is that subjects are observed in a completely unchanged environment.
- Under this approach surveys, observations and case studies are mainly used to collect the data and to understand the specific set of variables.

Example: A restaurant planned to start to serve continental food to its customers. Therefore, to learn about the choice of people in flavour and taste, observation method to learn about what kind of spreads, herbs and meat preferred by people.

4. Analytical

- This involves the use of the already available information.
- Here the researcher in an attempt to understand the complex problem set, studies and analyses the available with data.

- ❖ It majorly concerns the cause and effect relationship.
- ❖ The scientifically based problem-solving approaches mainly use this strategy.

Example: Examining the fluctuations of U.S. International trade balance during 1947–1995; while explaining why and how U.S. trade balance move in a particular way over time.

4. Action

- ❖ This strategy aims at finding solutions to an immediate problem.
- ❖ It is generally applied by an agency, company or by the government in order to address a particular problem and find possible solutions to it.

Example- Structured homework protocols for increasing student achievement.

5. Basic

- ❖ According to this strategy no generalizations are made in order to understand the subject in a better and precise way.
- ❖ Thus, it involves investigation and the analysis of a phenomenon.
- ❖ Although their findings are not directly applicable in the real world, they work towards enhancing the knowledge base.

Example- A study looking at how alcohol consumption impacts the brain.

6. Critical

- ❖ It works towards analysing the claims regarding a particular society.

Example- A Researcher design focus on any conclusion or the theory made regarding a particular society or the culture and test it empirically through a survey or experiment.

7. Interpretive

- ❖ This strategy is similar to the qualitative research strategy.
- ❖ However, rather than using hypothesis testing, interpretation is done through the sense-making process.
- ❖ In simple terms, this strategy uses human experience in order to understand the phenomena.

Example- Influence of dating violence on Adolescent Girls Educational Experiences

7. Exploratory

- ❖ It is mainly used to gain insights into the problem or regarding certain situations but does works towards providing the solution to the researcher problem.
- ❖ This research strategy is generally undertaken when there is very few or no earlier study on the research topic.

Example- A study into the implications of COVID-19 pandemic into the global economy

8. Predictive

- It deals with developing and understanding the future of the research problem and has its foundation based on the probability.
- This is generally very popular among companies and organizations.

Example- A Researcher might collect high school data such as grades, extracurricular activities, advanced courses taken and standardized test scores in order to predict such college success measures as grade- point average at graduation and awards received.

Extent of Researcher Interference

- The extent of interference by the researcher with the normal flow of work at the workplace has a direct bearing on whether the study undertaken is Causal or Correlational.
- A *Correlational study* is conducted in the natural environment of the organization with minimum interference by the researcher with the normal flow of work.
- Studies conducted to establish Cause and effect relationships, the researcher tries to manipulate certain variables to study the effects of such manipulation on the dependent variable of interest.
- In other words, The researcher deliberately changes certain variables in the setting and interferes with the events as they normally occur in the organization.

Minimal Interference

Example - Send questionnaires for not to interfere the daily work.

Moderate Interference

Example- To measure the stress.

Excessive Interference

Example- Design the measurement in details and the researcher has intervened maximally with the normal setting.

Study Setting

Contrived (Artificial) & Non Contrived(Usual):

Correlational studies are conducted in non- contrived settings (normal settings) whereas most Causal studies are done in contrived settings.

- Correlational studies done in organizations are also called FIELD STUDIES.
- Studies conducted to establish Cause and Effect Relationship using the Same

Natural Environment in which employees normally function are called FIELD EXPERIMENTS.

- Experiments done to establish Cause and Effect Relationship in a Contrived Environment and Strictly Controlled are called LAB EXPERIMENTS.

Unit of Analysis

The unit of analysis refers to the Level of Aggregation of the data collected during the subsequent data analysis.

- Individual
- Dyads
- Groups
- Organizations
- Cultures

Individual

If the researcher focuses on how to raise the motivational levels of employees, then we are interested in individual employees in the organization. Here the unit of analysis is the individual (the data will be gathered from each individual).

Dyads

If the researcher is interested in studying two-person interaction then several two-person groups also known as Dyads, will become the unit of analysis.

For example

Analysis of husband-wife and supervisor-subordinate relationships at the work place.

Time Horizon

(Cross-Sectional V/s Longitudinal Studies)

Cross Sectional Studies:

A Study can be done in which data are gathered just once perhaps over a period of days or weeks or months in order to answer a research question.

EXAMPLE:

Data were collected from stock brokers between April and June of last year to study their concerns in a turbulent stock market.

Data has to be collected at one point in time it is Cross-Sectional design.

Longitudinal Studies:

- Studying people or phenomena at more than one point in time in order to answer the research question.

- ❖ Because data are gathered at two or more different points in time, the study is not cross-sectional kind, but is carried longitudinally across a period of time.

Example:

A Marketing Manager is interested in tracing the pattern of sales on a particular product in four different regions of the country on a Quarterly Basis for the next 2 years.

Since the data are collected several times to answer the same issue the study falls under the longitudinal category.

Sources of Errors while Specifying Research Design

1. Population Specification

Population specification errors occur when the researcher does not understand who they should survey. This can be tricky because there are multiple people who might consume the product, but only one who purchases it, or they may miss a segment looking to purchase in the future.

2. Sampling And Sample Frame Errors

Survey sampling and sample frame errors occur when the wrong subpopulation is used to select a sample or because of variation in the number or representativeness of the sample that responds but the resulting sample is not representative of the population concern.

3. Selection

Selection error is the sampling error for a sample selected by a non-probability method. When respondents choose to self-participate in a study and only those interested respond, you can end up with selection error because there may already be an inherent bias. This can also occur when respondents who are not relevant to the study participate or when there's bias in the way participants are put into groups.

4. Non-Responsive

Non-response error can exist when an obtained sample differs from the original selected sample. This may occur because either the potential respondent was not contacted or they refused to respond. The key factor is the absence of data rather than inaccurate data.

5. Measurement

Measurement error is generated by the measurement process itself, and represents the difference between the information generated and the information wanted by the researcher. Generally, there is always some small level of measurement error due to uncontrollable factors.

Internal and External Validity:

Internal Validity

- Internal Validity is the extent to which a study establishes a trustworthy Cause and Effect relationship between a treatment and an outcome.
- Internal validity also reflects that a given study makes it possible to eliminate alternative explanations for a finding.
- Internal validity depends largely on the procedures of a study and how rigorously it is performed.

Internal validity is not a "Yes or No" type of concept. Instead, we consider how confident we can be with the findings of a study, based on whether it avoids traps that may make the findings questionable.

External Validity

- External validity refers to how well the outcome of a study can be expected to apply to other settings.
- In the other words, this type of validity refers to how generalizable the findings are
- For instance,
- Do the findings apply to other people, settings, situations and time periods.
- Ecological validity, an aspect of external validity, refers to whether a study's findings can be generalized to the real world.

Differences Between Internal Validity and External Validity

INTERNAL VALIDITY	EXTERNAL VALIDITY
Refers to the structure of a study and its variables.	Refers to how universal the results are.
Conclusions are warranted.	Findings can be generalized.
Controls extraneous variables.	Outcomes apply to practical situations.
Eliminates alternative explanations.	Results apply to the world at large.
Focus on accuracy and strong research methods.	Results can be translated into another context.

Choosing Right Research Design

A Research Design is a strategy for **answering your Research Question** using empirical data. Creating a research design means making decisions about:

- Your overall aims and approach.
- The type of research design you will use.
- Your sampling methods or criteria for selecting subjects.
- The procedures you will follow to collect data.
- Your data analysis methods.

- ❖ Consider your aims and approach.
- ❖ Choose a type of Research Design.

Triangulation

- ❖ Triangulation is a methodological research technique supposed to **support both the validation and the competences of research findings** in such a complex environment.
- ❖ Triangulation as a methodological technique refers to **the use of multi-method approach** in which different investigative methods are applied to the same phenomenon inorder to pinpoint the phenomenon, to improve the accuracy of the observations and to facilitate validation of collected data.

Importance in Behavioural Research

- ❖ Triangulation helps validate research findings by checking that different methods or different observers of the same phenomenon produce the same results.
- ❖ Researchers use triangulation for a more holistic perspective on a specific research question.
- ❖ Triangulation is also helpful for enhancing credibility and validity.
- ❖ Can help ensure that fundamental biases arising from the use of a single method or a single observer are overcome.

Types of Triangulation

For achieving a "Balanced picture of the situation" (Altricher , et al., 2008) Triangulation draws on using a combination of several research options. Methods, theories, sources of data and multiple numbers of researchers are all potential methodological alternatives techniques in the same phenomenon, while *Denzin(1978)* suggested *four different types, Guion (2002)* stressed the environmental factor as the *fifth type* of triangulation:

Data Triangulation

Involves time, space and persons.

Example

- ❖ To understand the motivations behind cooperative behavior, you compile and analyze data from a sample of 130 US college students over a period of 8 months. Then, you repeat the experiment with comparable samples in different regions worldwide.
- ❖ You collect data from participants in Germany and Japan to test your hypothesis using a wider sample.
- ❖ When you collect data from different samples, places or times, your results are more likely to be generalizable to other situations.

Investigator Triangulation

Involves **multiple researchers** in an investigation.

Example

- For your behavioral data, you involve multiple observers to code your participants behaviors. You provide them with training sessions and a manual to follow closely so they code behaviors the exact same way.
- They review video recordings of your participants playing team games in pairs and analyze and note down any cooperative behaviors. You check that their code sheets line up with each other to ensure high interater reliability.
- They also recalibrate the way they code behaviors intermittently for consistency
- Helps you reduce the risk of observer bias and other experimenter biases.

Theory Triangulation

Involves using more than one theoretical position in the interpretation of the phenomenon.

Example

- You believe there are two main competing motivational theories for why people behave cooperatively.
- People cooperate for a sense of reward: they cooperate to feel good.
- People cooperate to avoid guilt: they cooperate to avoid feeling bad.
- By gathering MR data, you can investigate whether there's more brain activity in reward-related or in guilt-averse brain areas when people cooperate.
- Testing competing hypotheses is one way to perform theory triangulation. Using theory triangulation may help you understand a research problem from different perspectives or reconcile contradictions in your data.

Methodological Triangulation

Involves using of more than one method for gathering data (interviews, questionnaries, documents, etc.).

Example

- In your study, you use behavioral, survey, and neural data to get a complete picture of what motivates people to behave cooperatively.
- You recruit participants to perform team games in a behavioral controlled lab experiment and record observations. You also administer a survey to gather data about cooperation in their daily lives. Finally, you perform fMRI scans to assess neural mechanisms of cooperation.
- Methodological triangulation is useful because you avoid the flaws and bias that come with reliance on a single research technique.

Environmental Triangulation

Involves the use of different locations, settings and other key factors related to the environment in which the study took place, such as time of the day, day of the week or season of the year.

Reference

R/R/PRA For Research and Extension in Agriculture Dr. Sabarathnam published on 1996 ISBN

https://www.upgrad.com/blog/5-types-of-research-design-elements-and-characteristics/

https://www.projectguru.in/how-to-formulate-a-research-strategy/

https://conjointly.com/kb/unit-of-analysis/

https://www.qualtrics.com/blog/5-common-errors-in-the-research-process/

https://www.verywellmind.com/internal-and-external-validity-4584479#:~:text=What%20is%20Internal%20Validity%3F,alternative%20explanations%20for%20a%20finding

https://www.scribbr.com/methodology/triangulation/

Questions

1. What is Research Design and it's elements?
2. What is the main purpose of Research Strategies?
3. What are the types of Research Strategies and explain them?
4. What is Correlational Study?
5. Define Field experiments and Lab experiments.
6. What are five Units of Analysis.
7. How Time Horizon differ in Studies?
8. What are the Sources of Errors while specifying Research Design?
9. Difference between Internal Validity and External Validity?
10. How to choose right Research Design?
11. Define Triangulation.
12. How Triangulation important in behavioural research?
13. What are the types of Triangulation?
14. Explain the types of Triangulation.
15. Define correlational.

Fill-Ups

1. **Research Design** is a blueprint for data collection, measurement and analysis.
2. A research strategy refers to **the step-by-step plan of action** that gives direction to the researcher's thought process.
3. **Descriptive** involves observing and describing the behaviour patterns of either an individual, community or any group.
4. **Action** aims at finding solutions to an immediate problem.
5. **Predictive** deals with developing and understanding the future of the research problem and has its foundation based on the probability.
6. The extent of interference by the researcher with the normal flow of work at the workplace has a direct bearing on whether the study undertaken is **Causal or Correlational.**
7. The **unit of analysis** refers to the Level of Aggregation of the data collected during the subsequent data analysis.
8. If the researcher is interested in studying two-person interaction then several two-person groups also known as **Dyads**, will become the unit of analysis.
9. **Non-response** error can exist when an obtained sample differs from the original selected sample.
10. **Internal validity** is not a "Yes or No" type of concept.
11. **Triangulation** is also helpful for enhancing credibility and validity.

Chapter 19

Research Methods and Instruments

Designing Research Instruments

Research Instrument

1. These are the fact finding strategies
2. They are the tools for data collection
3. They include Questionnaire, Interview and Observation

Characteristics of a Good Research Instrument

- It must be valid and reliable
- It must be based upon the conceptual framework
- It must gather data relevant to research topic
- It must gather data would test the hypotheses or answer the questions under investigation
- It should be free from all kinds of bias
- It must be accompanied by a good cover letter
- It must be accompanied, if possible, by a letter of recommendation from a sponsor.

Collection of Data through Questionnaires

- Sent (usually by post) to the persons concerned with a request to answer the questions and return the questionnaire
- Mailed to respondents who are expected to read and understand the questions
- The respondents have to answer the questions on their own Before using this method, it is always advisable to conduct 'pilot study' (pilot survey) for testing the questionnaires.

Main Aspect of Questionnaire

- The general form
- Question sequence
- Question formulation and wording

The General Form

Structured Questionnaire

- Definite, concrete and pre-determined questions
- Same wording and in the same order to all respondents
- Either closed (i.e., Of the type 'yes' or 'no') or open (i.e., Inviting free response) but should be stated in advance and not constructed during questioning.

Unstructure Questionnaire

- The interviewer is provided with a general guide on the type of information to be obtained, but the exact question formulation is largely his own responsibility
- The replies are to be taken down in the respondent's own words to the extent possible; in some situation tape recorders may be used to achieve this goal.

Question Sequence

- A proper sequence of questions reduces considerably the chances of individual questions being misunderstood
- Must be clear and smoothly-moving
- Relation of one question to another should be readily apparent to the respondent
- Easier question should be at the start of the questionnaire
- General to specific questions should be the sequence of questions.

Question Formulation and Wording

- Should also be impartial in order not to give a biased picture of the true state of affairs
- Should be easily understood
- Should be simple i.e., Should convey only one thought at a time
- Should be concrete and should conform as much as possible to the respondent's way of thinking.

Two principal forms of questions:

- Multiple choice question
    - Respondent selects one of the alternative possible answers put to him
- Most amenable to statistical analysis
    - Open-end question
- Respondent has to supply the answer in his own words.

Types of questionnaires:

- Structured questionnaires
- Unstructured questionnaires

- Semi-structured Questionnaires

Essentials of a Good Questionnaire:

- Short and simple
- Logical sequence
- Personal and intimate questions should be left to the end
- Technical terms and vague expressions should be avoided
- Adequate space for answers
- There should always be provision for indications of uncertainty

Merits of Questionnaires:

- Low cost
- It is free from the bias of the interviewer; answers are in respondents' own words
- Respondents have adequate time to give well thought out answers
- Respondents, who are not easily approachable, can also be reached conveniently
- Large samples can be made.

Demerits of Questionnaires:

- Low rate of return of the duly filled in questionnaires; bias due to no-response is often indeterminate
- The control over questionnaire may be lost once it is sent
- There is also the possibility of ambiguous replies or omission of replies altogether to certain questions; interpretation of omissions is difficult
- It is difficult to know whether willing respondents are truly representative
- This method is likely to be the slowest of all.

Interview

This is a method in which the investigator and the respondent meet and questions raised are answered and recorded.

- This method is adopted when personal opinion or view point are to be gathered as a part of the data.
- One of the very old methods of compiling data.
- This method has to be necessarily adopted whenever details regarding any confidential matter should be collected or when authentic information about anything is to be obtained.

Types Of Interviews

1. Direct Interview:

- In this type of interview, the interviewer and the interviewee meet personally either with prior appointment or not.
- This enables the respondent to be ready with necessary details or data. ex: The chairman of ISRO is to be interviewed
- In some cases, the respondent is contacted and is opinion collected.
- In such cases, there is no possibilities for briefing him about the interview. ex: TV crews meet the people coming out a theatre screening a new film & ask them about the film.

Structured interview:

- In this type of interview, the person collecting the information decides in advance, the nature and scope, questions to be asked the person to be contacted, etc.
- At the time of interview, no deviation is made from the questions to be asked, the details to be quoted, the instance to be recalled, etc. (Ex: Journalist to interview the finance minister after the presentation of budget).

Unstructured interview:

- Interview is conducted on a spot without any preparations or advance information to the respondent. ex: An organization producing a new health drink wants to know the opinion of the people directly
- As such interviews take place with people not informed in advance except the location nothing can be decided by the organization.
- Such interviews used when an opinion poll is conducted. ex: After the budget is presented in the parliament, public opinion is collected on the announcements in the budget.

Non -Directive interview:

- Unstructured type of interview.
- Interviewer encourages the respondents to say whatever he likes and feels on a subject matter.
- There may not be much of questioning taking place.

Telephonic Interview

- Essentially a type of direct interview but there is no scope for physical presence of both parties to the interview.
- Used when the informant and the interviewer are geographically separated.

Merits of Interview

- Non response is low
- Enables personal contact with which factual information could be obtained.
- Possible to obtain personal information
- Interviewee can guide the respondents to understand the questions
- Interviewer has the flexibility to handle the interview to collect relevant information.

Demerits of Interview

- Costly method of data collection
- Possible for a respondent to conceal his real opinion & so genuine data may not be available.
- Requires an elaborate organizational arrangement.
- Interviewer may also be biased & thereby his bias may influence the interview outcome.

Collection of Data through Schedules

- Schedules (proforma containing a set of questions) are being filled in by the enumerators.
- Enumerators along with schedules, go to respondents, put to them the questions and record the replies
- In certain situations, schedules may be handed over to respondents and enumerators may help them in recording their answers
- Ex. Population census all over the world is conducted through this method.

Questionnaire	Schedule
Questionnaires do not involve person to person interview.	Schedules are accompanied by person to person interviews.
Questionnaires mainly involve questions.	Schedules involves statements, blanks, questions, etc.
Questionnaires do not involve participation of the collector.	Schedules involve active participation of the interviewer.
Needs to be filled by respondents.	Can be filled by enumerators or respondents anyone.
Response rate remains low.	Response rate is high.
Overall cost is low	The overall cost is expensive.
Can't be done when respondents are illiterate.	Can be done when respondents are illiterate.

Test Definition

- A test is a set of questions, each of which has a correct answer, to which the people respond

- A test is an organized succession of stimuli, designed to measure quantitatively some mental process, trait or characteristics

1. Knowledge tests:

- Knowledge as a body of understood information possessed by an individual or a culture
- Influences covert and overt behavior of a person

Steps in Knowledge test:

- Item collection
  - Initial selection of items should promote thinking rather than only memorization
  - Should differentiate between well informed respondents from poorly informed one
  - The items should be in objective form and yes/no, true/false, multiple choice etc.
- Screening out of statements after consultation with experts
- Item analysis:

Based on,

i. Item Difficulty

ii. Item discrimination

iii. Point biserial correlation co-efficient.

Difficulty index

- The difficulty index indicates the extent to which an item is difficult
- An item should not be too easy that all persons can pass it, nor should be too difficult that none can pass it

Discrimination index (DI)

- The function of discrimination index is to find out whether an item really discriminates from a informed respondent to a poorly informed respondent
- Score of total respondents should be rank in descending order

Point biserial correlation coefficient

- For establishing the internal consistency of each item, point biserial correlation coefficient was estimated

2. Behaviour performance tests

- Behavioral assessment is a structured study and analysis of a person's behavior using various methods like interviews, direct supervision, and self-observation

- It refers to the process of observing, explaining, and predicting human behavior

Steps in behavior performance test:

1.Competency-Tool Mapping

- Competency-tool mapping involves creating an organization specific competency framework to identify relevant indicators of behavior& mapping them on available tools.

2.Tool Creation

- Tools and questions are created based on specific job functions.
- They emulate the job's realities, posing questions that reflect actual workplace situations.

3. Content Validation

- Content validation is vital to obtain optimal results.
- Multiple iterations of the content in consultation with subject matter experts and the organization are the best practice to reflect the real essence of the organizational environment.

4.Test Administering & Report Generation

- For online behavioral tests, organizations can opt for virtual assessment centers. They are easy to roll out, highly time and cost-effective, easily scalable and generate instant reports
- Offline tools often involve one or more assessors observing the candidates at a physical or blended assessment center. The reports typically take longer because the assessors need to cull the results together.

5. Data Analysis & Individual Development Plans

- The reports should provide a detailed behavioral profile of the test-taker
- Data analysis can help organizations understand the key skill gaps and training needs at a group level
- Individual development plans are customized for each test-taker to help them understand and work on their development areas.

Scaling

Scaling technique is a method of placing respondents in continuation of gradual change in the pre-assigned values, symbols or numbers based on the features of a particular object as per the defined rules. All the scaling techniques are based on four pillars, i.e., order, description, distance and origin.

Scaling Types:

- Primary Scaling Techniques

- ▷ Nominal Scale
- ▷ Ordinal Scale
- ▷ Interval Scale
- ▷ Ratio Scale

❖ Other Scaling Techniques

- ▷ Comparative Scales
- ▷ Non-Comparative Scales

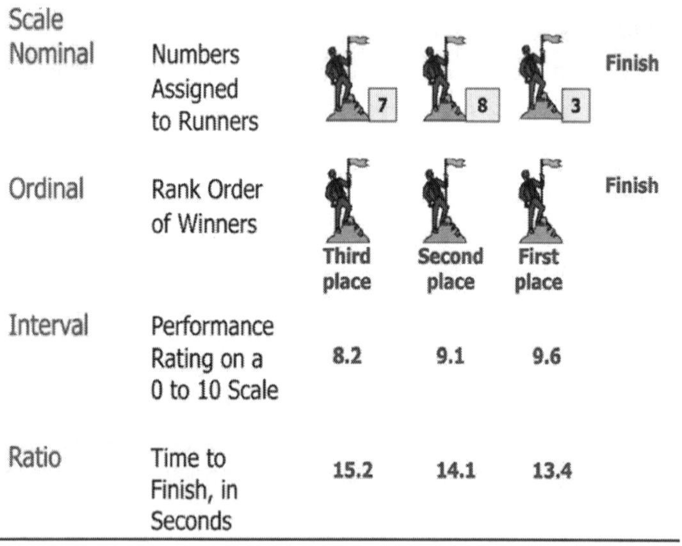

Comparative Scales:

For comparing two or more variables, a comparative scale is used by the respondents.

❖ Paired Comparison

- ▷ A paired comparison symbolizes two variables from which the respondent needs to select one. This technique is mainly used at the time of product testing, to facilitate the consumers with a comparative analysis of the two major products in the market.

❖ Rank Order

- ▷ In rank order scaling the respondent needs to rank or arrange the given objects according to his or her preference.

❖ Constant Sum

- ▷ It is a scaling technique where a continual sum of units like dollars, points, chits, chips, etc. is given to the features, attributes and importance of a particular product or service by the respondents.

Non-comparative scale:

- It is used to analyze the performance of an individual product or object on different parameters. Following are some of its most common types:

Continuous rating scale:

- It is a graphical rating scale where the respondents are free to place the object at a position of their choice. It is done by selecting and marking a point along the vertical or horizontal line which ranges between two extreme criteria.

Itemized scale

- is another essential technique under the non-comparative scales. It emphasizes on choosing a particular category among the various given categories by the respondents. Each class is briefly defined by the researchers to facilitate such selection.

Likert Scale

- In the Likert scale, the researcher provides some statements and ask the respondents to mark their level of agreement or disagreement over these statements by selecting any one of the options from the five given alternatives.

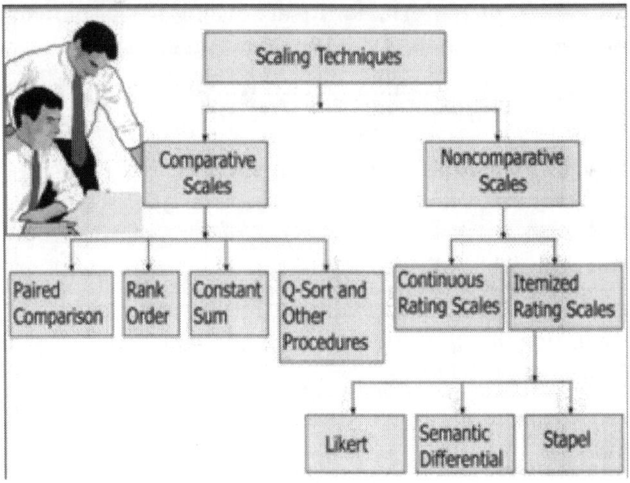

Participatory Rural Appraisal Tools and Techniques:

- Participatory extension is a learning approach for strengthening individual and organizational capabilities of rural people and their livelihoods to enable them to better cope with development in a self-reliant way.
- It is a multi-directional communication process between and among extension staff and farmers.

Participatory Tools and Techniques:

Space Related Tools

- ❖ Social map
- ❖ Resource map
- ❖ Mobility map
- ❖ Transect walk

Time Related Tools

- ❖ Time line
- ❖ Trend Analysis
- ❖ Seasonal Diagram

Relational methods

- ❖ Venn diagram
- ❖ Matrix Scoring/ Ranking method

Social Map:

- ❖ Depict the habitation pattern, Nature of housing, Social-infrastructure: Roads, Drainage systems, schools, drinking water facilities, etc.
- ❖ Made by people not by experts, not drawn based on any scales.
- ❖ High authenticity.
- ❖ Effective: small & medium sized localities.

Resource Map:

- ❖ Most commonly used PRA
- ❖ Focuses on the natural resources in the locality and depicts land, hills, rivers, fields, vegetation, soil fertility, erosion, land use pattern, irrigation, watershed development etc.
- ❖ Not done by expert but only by the local people. as it will be more accurate and detail.

Application:

- ❖ Ask them to use various symbols and colors
- ❖ Ask them to discuss about the problems and opportunities regarding the objective.
- ❖ Make everyone involve.

Mobility Map:

- ❖ Used to explore the movement pattern of an individual, a group, or the community.
- ❖ Focus on where people go and what for.

- And also on frequency, distant, importance of the place.
- Application
- Understand mobility pattern of the local people
- Gender issues related to mobility pattern
- Evaluation of impacts in some interventions
- Planning interventions.

Transect Walk:

- Used to explore spatial realities
- Different from resource map despite areas of overlap.
- Resource map provides a bird's- eye view whereas transect gives a cross-sectional view of the locality

Application:

- Appraisal of natural resources in terms of status, problems and potential.
- Verification of issues aroused in other PRA practices.
- Planning various interventions.

Time Line:

- An aggregate of various landmark events in their community as perceived by the local people.
- The Time Line exercise provides an ideal opportunity to capture the chronology of events in an individual or community's history.
- Time Line represents important moments as perceived and recalled by the people involved.
- This tool can serve as an effective rapport builder.

Trend Analysis:

- A before and after type of analysis that identifies the changes that have taken place across certain time milestones.
- A trend analysis captures changes and trends related to certain variables over different spans of time.
- Crop yield, livestock population, rainfall etc...

Seasonal Mapping:

- An image of people's livelihood conditions over a period of time, the communities' notion of an annual cycle.
- It is used to understand changes in livelihood conditions, income availability, diseases, expenditure, and other entitlements of the poor, across different periods of the year.

Venn Diagram:

- Venn diagrams provide a visual representation of the relationships and linkages between people and institutions.
- Circles of different sizes are allocated to different institutions, groups, departments, or programmes, based on their importance.
- Size – importance of the institution or individual.
- Distance between circles – degree of Influence.

Matrix Scoring Ranking:

- Matrix scoring/ranking is a preferences analysis used for comparing different elements against a range of criteria.
- While simple ranking can be used as an indicator, more complex matrix and scoring techniques enables a criteria based in-depth analysis.

Projective and Non-Projective Techniques:

Projective Techniques:

- Projective test in which the individual offer response to ambiguous sense, word or image thus type of text emerged from the psychology school through which suggested that people have unconscious thought or urges.
- Types of projective tests
- Rorschach inkblot test
- The four-picture test
- T.A.T
- H.T.P
- Word Association Test
- RISB

Non-Projective Techniques

- Non-projective tests are psychological tests that measure an individual's characteristics in a way that is independent of rater bias or the individual's own belief.
- **Two types**- Interview, Questionnaire

References

https://www.google.com/search?q=research+design+in+research+methodology&oq=RESEARCH+DESIGN+&aqs=chrome.1.69i57j0i512j0i433i512j0i51213j46i199i465i512j0i51213.6962j0j15&sourceid=chrome&ie=UTF-8

https://www.advanceinnovationgroup.com/blog/experimental-designs-as-variance-control

Research methodology – a step by step guide for beginners by Ranjithkumar

Research methodology by Prakash S.Kattimani, BasavarajBiradar

Questions

1. What is designing research instruments?
2. What all are the characteristics of good research instruments?
3. How to collect the data through questionnaires ?
4. Explain the types of questionnaires ?
5. Define- interview
6. What all are the types of interview?
7. Difference between questionnaire and schedule
8. Define-test
9. Explain scaling types
10. define - PRA and tools of PRA

Fill Ups

1. Designing research instruments includes _____
2. _____ is One of the Characteristics of a Good Research Instrument
3. Main aspect of questionnaire is ____
4. Definite, concrete and pre-determined questions are called as _____
5. Merits of Questionnaire is____
6. ______is a method in which the investigator and the respondent meet and questions.
7. _____ is a type of direct interview but there is no scope for physical presence of both parties to the interview.
8. Population census all over the world is conducted through ____ method.
9. A ______is a set of questions, each of which has a correct answer, to which the people respond.
10. ______is a method of placing respondents in continuation of gradual change in the pre-assigned values, symbols or numbers based on the features of a particular object as per the defined rules.
11. _______is a learning approach for strengthening individual and organizational capabilities of rural people and their livelihoods to enable them to better cope with development in a self -reliant way.
12. ______is Used to explore the movement pattern of an individual, a group, or the community.

13. ______ is provide a visual representation of the relationships and linkages between people and institutions.
14. In rank order scaling the respondent needs to ______the given objects according to his or her preference.
15. _____ indicates the extent to which an item is difficult.

Chapter 20

Sampling Concepts and Probability Sampling Techniques

Sampling is a technique of selecting individual members or a subset of the population to make statistical inferences from them and estimate characteristics of the whole population. Different sampling methods are widely used by researchers in market research so that they do not need to research the entire population to collect actionable insights.

It is also a time-convenient and a cost-effective method and hence forms the basis of any research design. Sampling techniques can be used in a research survey software for optimum derivation.

For example, if a drug manufacturer would like to research the adverse side effects of a drug on the country's population, it is almost impossible to conduct a research study that involves everyone. In this case, the researcher decides a sample of people from each demographic and then researches them, giving him/ her indicative feedback on the drug's behaviour.

Population

The population can be defined in terms of geographical location, age, income, and many other.

- ❖ Population is the aggregation of all the units in which a researcher is interested.
- ❖ It refers to the group of people, items or units under investigation and includes every individual.
- ❖ The population is the entire group that you want to draw conclusions about.
- ❖ The sample is the specific group of individuals that you will collect data from.

It can be very broad or quite narrow: maybe you want to make inferences about the whole adult population of your country; maybe your research focuses on customers of a certain company, patients with a specific health condition, or students in a single school.

It is important to carefully define your target population according to the purpose and practicalities of your project.

If the population is very large, demographically mixed, and geographically dispersed, it might be difficult to gain access to a representative sample.

Sample

- A sample is a finite part of statistical population whose properties are studies to gain information about the whole (Webster, 1985).
- It is a miniature representation of and selected from a larger group or aggregate.
- In other words, the sample provides a specimen picture of a larger whole.

Element: it is the basic unit that represents whatever is being sampled and from which survey data are to be gathered.

Sampling unit: It is one of the units selected for the purpose of sampling

Sampling frame: The sampling frame is the actual list of individuals that the sample will be drawn from. Ideally, it should include the entire target population (and nobody who is not part of that population).

Sample size: The number of individuals you should include in your sample depends on various factors, including the size and variability of the population and your research design. There are different sample size calculators and formulas depending on what you want to achieve with statistical analysis.

Sampling strategies for qualitative research:

- The two most popular sampling techniques are
  - Purposeful sampling
  - Convenience sampling

Purposeful sampling

- Technique that qualitative researchers use to recruit participants who can provide in depth and detailed information about phenomenon under investigation.
- It is highly subjective and determined by the qualitative researcher generating the qualifying criteria each participant must meet to be considered for the research study.

Convenience sampling

- Technique that qualitative researchers use to recruit participants who are easily accessible and convenient to the researchers.
- Often time this may include utilizing geographic location and resources that make participant recruitment convenient.

Sampling strategies for quantitative research

- ❖ Quantitative research refers to the analysis wherein mathematical, statistical, or computational method is used for studying the measurable or quantifiable dataset.
- ❖ The core purpose of quantitative research is the generalization of a phenomenon or an opinion. This involves collecting and gathering information from a small group out of a population or universe.
- ❖ It includes probability sampling techniques in the majority of cases and also includes non-probability sampling technique in some cases.

Principles of sampling

- ❖ **1st Principle:** In a majority of cases of sampling there will be difference between the sample statistics and the true population mean, which is attributable to the selection of the units in the sample.
- ❖ **2nd Principle:** The greater the sample size, the more accurate will be the estimate of the true population mean.
- ❖ **3rd principle:** The greater the difference in the variable under study in a population for a given sample size, the greater will be the difference between the sample statistics and the true population mean.

Factors affecting the inferences drawn from a sample

- ❖ The size of the sample: findings based upon larger samples have more certainty than those based on smaller ones. As a rule, the larger the sample size, the more accurate will be the finding.
- ❖ The extent of variation in the sampling population: the greater the variation in the study population with respect to the characteristics under study for given sample size, the greater will be the uncertainty. (In technical terms, the greater the standard deviation, the higher will be the standard error for a given sample size in your estimates).

Types of sampling: sampling methods

Sampling is two types – probability sampling and non-probability sampling. Let's take a closer look at these two methods of sampling.

1. Probability sampling: Probability sampling is a sampling technique where a researcher sets a selection of a few criteria and chooses members of a population randomly. All the members have an equal opportunity to be a part of the sample with this selection parameter.

2. Non-probability sampling: In non-probability sampling, the researcher chooses members for research at random. This sampling method is not a fixed or predefined selection process. This makes it difficult for all elements of a population to have equal opportunities to be included in a sample.

Methods of drawing random sample

- The fishbowl draw – if your total population is small, an easy procedure is to number each element using separate slips of paper for each element, put all slips into a box, and then pick them out one by one without looking, until the number of slips selected equals the sample size you decide upon. This method is used in some lotteries.
- Computer program – there are number of programs that can help you to select a random sample.
- A table of random numbers

Sampling with and without replacement

- There are two ways of selecting a random sample
- If we draw a ball from an urn containing balls numbered 1 to N, we have the choice of replacing or not replacing the ball into the urn before a second ball is drawn
- In the first case the particular ball can be drawn again and again, whereas in the second case it can only be selected once. Sampling where each unit of population may be chosen more than once is called as sampling with replacement, while if each unit cannot be chosen more than once it is called as sampling without replacement.

Probability sampling

- For a sampling design to be called a random or probability sample, it is imperative that each element in the population has an equal and independent chance of selection in the sample.
- Equal implies that probability of selection of each element in the population is the same, i.e. the choice of an element in the sample is not influenced by other considerations such as personal preference.

There are four types of probability sampling techniques

- Simple random sampling: One of the best probability sampling techniques that helps in saving time and resources, is the Simple Random Sampling method. It is a reliable method of obtaining information where every single member of a population is chosen randomly, merely by chance. Each individual has the same probability of being chosen to be a part of a sample.
  - Identify number of elements or sampling units in the population
  - Decide on the sample size (n)
  - Select (n) using either the fish bowl draw, the table of random numbers or a computer program.

Systematic sampling

- The sample is chosen by selecting a random starting point and then picking each ith element in succession from the sampling frame.

- The sampling interval i, is determined by dividing the population size N by the sample size n and rounding to the nearest integer.
- For example, if there were 30 owners of new washing machine and a sample of 10 is to be desired, the sampling interval i is 3
- The researcher then selects a number between 1 and 30. if, for example number (k) is chosen by the researcher the sample will consist of elements 3(k), (k+i)6, (k+2i) 9......k+(sample size-1)i.

Stratified random sampling

- In this researcher attempts to stratify the population in such a way that the population within a stratum is homogeneous with respect to the characteristics on the basis of which it is being stratified.
- Once the sampling population has been separated into non-overlapping groups you select the required number of elements from each stratum using the simple random sampling technique.

Types

Proportionate stratified sampling - The number of elements from each stratum in relation to its proportion in the total population is selected.

Disproportionate stratified sampling - Consideration is not given to the size of the stratum

- Identify all elements or sampling units in the sampling population
- Decide upon the different strata (k) into which you want to stratify the population
- Place each element into the appropriate stratum
- Number every element in each stratum separately
- Decide the total sample size(n)
- Decide whether you want to select proportionate or disproportionate stratified sampling

Proportionate stratified sampling

- Determine the proportion of each stratum in the study population (p)
- Determine the number of elements to be selected from each stratum
- Select the required number of elements from each stratum with SRS technique
- As the sample selected is in proportion to the size of each stratum in the population, this method is called proportionate stratified sampling

Disproportionate stratified sampling

- Determine the number of elements to be selected from each stratum
- Select the required number of elements from each stratum with SRS technique.

- ❖ As this method does not take the size of the stratum into consideration in the selection of sample, it is called disproportionate stratified sampling.

Cluster sampling

- ❖ Simple random and stratified sampling techniques are based on the researcher's ability to identify each element in a population. It is easy to do this if the total sampling population is small, but if the population is large, as in the case of the city, state or country, it becomes difficult and expensive to identify each sampling unit. In such cases the use of cluster sampling is more appropriate.
- ❖ Cluster sampling is based on the ability of the researcher to divide the sampling population into groups, called clusters, and then to select elements within each cluster, using SRS technique.

Uses of probability sampling

There are multiple uses of probability sampling:

- ❖ Reduce Sample Bias: Using the probability sampling method, the bias in the sample derived from a population is negligible to non-existent. The selection of the sample mainly depicts the understanding and the inference of the researcher. Probability sampling leads to higher quality data collection as the sample appropriately represents the population.
- ❖ Diverse Population: When the population is vast and diverse, it is essential to have adequate representation so that the data is not skewed towards one demographic. For example, if Square would like to understand the people that could make their point-of sale devices, a survey conducted from a sample of people across the US from different industries and socio-economic backgrounds helps.
- ❖ Create an Accurate Sample: Probability sampling helps the researchers plan and create an accurate sample. This helps to obtain well-defined data.

Multiple Choice Questions

1. **Which of the following sampling methods involves dividing the population into subgroups or strata, then randomly selecting individuals from each stratum?**

   a) Stratified Random Sampling

   b) Snowball Sampling

   c) Quota Sampling

   d) Convenience Sampling

2. **Which of the following sampling techniques involves selecting individuals who are readily available without any consideration for their representativeness of the larger population?**

   a) Snowball Sampling

   b) Systematic Sampling

   c) Stratified Sampling

   d) Convenience Sampling

3. **Which of the following sampling methods is most likely to result in a biased sample?**

   a) Random sampling

   b) Stratified random sampling

   c) Convenience sampling

   d) Quota sampling

4. **A college principal conduct an ethnographic probe into the problems faced by tribal students. Which method of sampling will be most appropriate?**

   a) Random sampling

   b) Stratified sampling

   c) Cluster sampling

   d) Systematic sampling

5. **Snowball sampling technique falls under the category of:**

   a) Probability sampling

   b) Non-probability sampling

   c) Random sampling

   d) Quota sampling

6. **Snowball sampling is the process of selecting a sample using**

   a) Networks

   b) Groups

   c) Snowballs

   d) Computer Programs

7. **Which of the following sampling methods is based on probability?**

   a) Judgement sampling

   b) Stratified sampling

   c) Quota sampling

   d) Convenience sampling

8. For the selection of samples in research, the snowball sampling method employs

a) Field workers

b) Opinion leaders

c) Surveyors

d) Referrals

9. What do we say to all units aggregate that's about a study?

a) sample

b) unit

c) universe or population

d) frame

10. What refers to elements from where you choose the samples for the research?

a) infinite population

b) finite population

c) sampling population

d) target population

Questions

1. What do you mean by sampling? Discuss the factors that influence the decision to sample.
2. What is sampling distribution? Discuss the methods of drawing random samples.
3. Discuss the relative advantages and disadvantages of any two probability sampling methods with example.
4. Explain the importance of sampling in research. Discuss any two methods of sampling.
5. What are the requisites of a good sample? State the advantages of randomization in scientific research.
6. When is cluster sampling likely to be preferred most?
7. Why would a researcher choose stratified random sampling instead of simple random sampling?

Chapter 21

Non-Probability Sampling

Non-Probability Sampling

- Non-probability sampling is a method of selecting units from a population using a subjective (i.e. non-random) method.
- Since non-probability sampling does not require a complete survey frame, it is a fast, easy and inexpensive way of obtaining data.
- Probability sampling involves random selection, allowing you to make strong statistical inferences about the whole group. Non-probability sampling involves non-random selection based on convenience or other criteria, allowing you to easily collect data.

"Nonprobability sampling refers to any sampling method in which some elements of the population have no chance of selection (these are sometimes referred to as 'out of coverage'/'undercoverage'), or where the probability of selection can't be accurately determined." **Donald L. Rubin and Richard A. Hursh-Cesar (1978)**

"Nonprobability sampling is a sampling technique in which the researcher uses his judgment to select the subjects or units to be studied. Convenience sampling, judgment sampling, quota sampling, and snowball sampling are types of nonprobability sampling." **William G. Zikmund (2003)**

"Nonprobability sampling is any technique in which samples are selected in some way not suggested by probability theory. In nonprobability samples, some individuals have little or no chance of being selected, while others may be highly likely to be chosen." **Paul D. Leedy and Jeanne Ellis Ormrod (2015)**

"In non-probability sampling, the probability of each case being selected is unknown. Instead, we rely on our judgment or convenience. Non-probability samples are often used when we are interested in specific groups rather than in making generalizations about populations." **David A. de Vaus (2002)"**

"Non-probability sampling is a sampling technique where the samples are gathered in a process that does not give all the individuals in the population equal chances of being selected". **Kumar, R. (2014):**

Difference Between Non-Probability Sampling and Probability Sampling

Non probability sampling	*Probability sampling*
Sample selection based on the subjective judgment of the researcher.	The sample is selected at random.
Not everyone has an equal chance to participate	Everyone in the population has an equal chance of getting selected.
The researcher does not consider sampling bias.	Used when sampling bias has to be reduced.
Useful when the population has similar traits.	Useful when the population is diverse.
The sample does not accurately represent the population.	Used to create an accurate sample.
Finding respondents is easy.	Finding the right respondents is not easy.

When to use non-probability sampling?

- Use this type of sampling to indicate if a particular trait or characteristic exists in a population.
- Researchers widely use the non-probability sampling method when they aim at conducting qualitative research, pilot studies, or exploratory research.
- Researchers use it when they have limited time to conduct research or have budget constraints.
- When the researcher needs to observe whether a particular issue needs in-depth analysis, he applies this method.
- Use it when you do not intend to generate results that will generalize the entire population.

Types

- Reliance on Available Subjects
- Accidental sampling
- Judgemental sampling or Purposive sampling
- Quota sampling
- Expert sampling
    - Snowball sampling
    - Modal instance
    - Heterogenity sampling

Convenience Sampling

Convenience sampling (also called *accidental sampling* or *grab sampling*) is where you include people who are easy to reach.

For example, you could survey people from:

- ▷ Your workplace,
- ▷ Your school,
- ▷ A club you belong to,
- ▷ The local mall.

- ❖ It's relatively easy to get a sample.
- ❖ It's inexpensive, compared to other methods.
- ❖ Participants are readily available.
- ❖ No evidence for representativeness of population of interest

Reliance on Available Subjects

- ❖ Is a risky model that requires a great deal of caution on the part of the researcher.
- ❖ Since it entails sampling passersby or individuals with whom researchers randomly come into contact, it is sometimes referred to as a convenience sample because it does not allow the researcher to have any control over the representativeness of the sample.
- ❖ While this sampling method has drawbacks, it is useful if the researcher wants to study the characteristics of people passing by on a street corner at a certain point in time, especially if conducting such research would not be possible otherwise.

Though this method can be useful, the researcher will not be able to use the results from a convenience sample to generalize about a wider population.

Purposive or Judgemental Sampling

- ❖ Also known as judgmental, selective, or subjective sampling
- ❖ The sampling design is based on the judgement of the researcher as to who will provide the best information to succeed for the objectives study.
- ❖ The person conducting the research need to focus on those people with the same opinion to have the required information and be willing of sharing it.
- ❖ All a researcher must do is reject the individuals who do not fit a particular profile when creating the sample.
- ❖ Eg: A group of researchers is interested in learning if the reason why people wear eyeglasses is to read books.
- ❖ Common sense tells us that the efforts of the research group should be focused entirely on people that indeed wear eyeglasses. This process is judgment sampling in action

Advantages

- ❖ Consumes minimum time for execution: In this sampling approach, researcher expertise is important and there are no other barriers involved due to which selecting a sample becomes extremely convenient.
- ❖ Allows researchers to approach their target market directly: There are no criteria involved in selecting a sample except for the researcher's preferences. Due to this, he/she can communicate directly with the target audience of their choice and produce desired results.
- ❖ Almost real-time results: A quick poll or survey can be conducted with the sample using judgmental sampling since the members of the sample will possess appropriate knowledge and understanding of the subject

Quota Sampling

- ❖ Method in which researchers create a sample involving individuals that represent a population.
- ❖ Researchers choose these individuals according to specific traits or qualities.
- ❖ They decide and create quotas so that the research samples can be useful in collecting data.
- ❖ These samples can be generalized to the entire population. The final subset will be decided only according to the interviewer's or researcher's knowledge of the population.

Example

Situation 1:

A researcher wants to survey individuals about what smartphone brand they prefer to use. He/she considers a sample size of 500 respondents.

The researcher can divide the population by quotas:

Gender: 250 males and 250 females

Age: 100 respondents each between the ages of 16-20, 21-30, 31-40, 41-50, and 51+

Employment status: 350 employed and 150 unemployed people.

(Researchers apply further nested quotas. For eg, out of the 150 unemployed people, 100 must be students.)

Location: 50 responses per state

Advantages

- ❖ **Saves time:** Because of the involvement of a quota for sample creation, this sampling process is quick and straightforward.
- ❖ **Research convenience:** By using quota sampling and appropriate research questions, interpreting information and responses to the survey is a much convenient process for a researcher.

- **Accurate representation of the population of interest:** Researchers effectively represent a population using this sampling technique. There is no room for over-representation as this sampling technique helps researchers to study the population using specific quotas.
- **Saves money:** The budget required for executing this sampling method is minimalistic.

Expert Sampling

- Expert sampling is an example of purposive sampling
- The researcher here seeks for the consent of those that are expert or known expert in the area of study, and begin the process of collecting his information directly from individual or group of respondent.
- It also involves sample assembling of group of people that can demonstrate using their experience or those that specialised in part of the areas.
- Potential reason to use expert sampling:
  - Best way to elicit views of persons with specific expertise
  - Provide evidence for the validity of another sampling approach
  - Advantage: expert judgement supports the sampling
  - Problem: the "experts" may be wrong

Snowball Sampling (Chain-Referral Sampling)

- Is a design process of selection usually done by using, networks.
- It is useful when the researcher knows little about a group or organisation to study; contact with few individuals will direct him to other group.
- The selection of the study sample will be useful for communication aspect, in making decision or in difussion of knowledge to people.
- Snowball sampling method is purely based on referrals and that is how a researcher is able to generate a sample. Therefore, this method is also called the chain-referral sampling method
- This sampling technique can go on and on, just like a snowball increasing in size (in this case the sample size) till the time a researcher has enough data to analyze, to draw conclusive results that can help an organization make informed decisions.
- Can be extensively used for conducting qualitative research, with a population that is hard to locate.
- An individual, who is ready to share information, needs to know that the information will be used discreetly and this kind of trust is especially important in snowball sampling. For a participant to agree to identify themselves or their group, researchers first need to develop that kind of rapport with the participants.

"May consume more time than anticipated because of its nature."

Advantage:

- It's quicker to find samples
- Cost effective
- Sample hesitant subjects

Disadvantage

- Sampling bias and margin of error
- Lack of cooperation

Modal Instant Sampling

- In this type of sampling, we sample the most frequent cases
- It can also be seen as the one with the highest happening of value in a given distribution or the one with most characteristic incident.
- The term modal comes from the mode, which is the most common item in a set.
- it is only recommended as a method for **informal questionnaires or surveys.**
- For example, newscasters might interview a typical voter, or a typical resident, or even residents of a typical neighborhood

Heterogeneity Sampling

- It can be used if we want to include all the opinions or views and we are not going to consider about representing of these views per head.
- This sampling is also called **sampling of diversity** and is almost opposite with the modal instant sampling, the interest is to have a comprehensive variety of ideas, not to identify the modal instance or typical once.
- We determine that there is a ground of all possible ideas applicable to some topic and that we want to sample the population not the population of those people who have the ideas.
- **What the sample study need is ideas not people**
- **What should be the sample size?**

A **qualitative research** approach often has a **small sample size** (i.e. a small number of people researched) since each respondent will provide you with pages and pages of information in the form of interview answers or observations

Quantitative research relies heavily on the ability to gather data from a **large sample** and use it to explain a far larger population (this is called "generalisability").

Data collection for quantitative research involves large sample sizes but relatively basic data.

Misconceptions

- The sample size should be a particular proportion, say 5 per cent, of the population.
- The sample size should be about 2000.

However, no such thumb rule is adequate. In fact, the size of the sample is dependent on the level of accuracy required or how large an error is acceptable

The sample size should not be less than 30, if you want to employ parametric tests such as correlation, regression, t-test, ANOVA, MANOVA etc.

How to estimate sample size

1. Confidence intervals measure the degree of uncertainty or certainty in a sampling method and how much uncertainty there is with any particular statistic. In simple terms, the confidence interval tells you how confident you can be that the results from a study reflect what you would expect to find if it were possible to survey the entire population being studied.

Eg: if your confidence interval is 6 and 60% percent of your sample picks an answer, you can be confident that if you had asked the entire population, between 54% (60-6) and 66% (60+6) would have picked that answer.

2. Confidence level: It is expressed as a percentage and represents how often the percentage of the population who would pick an answer lies within the confidence interval.

Eg: a 99% confidence level means that should you repeat an experiment or survey over and over again, 99 percent of the time, your results will match the results you get from a population.

3. Standard deviation: Estimating how much the responses you receive will vary from each other and from the mean.

Higher the variability, greater the SD

Steps

- Determine the population
- Determine the confidence interval
- Determine the confidence level
- Determine SD
- Convert confidence level into Z score

Cochran formula

$$n_0 = \frac{Z^2pq}{e^2}$$

- e is the desired level of precision (i.e. the margin of error
- p is the (estimated) proportion of the population which has the attribute in question,
- q is 1 – p

For small sample

$$n = \frac{n_0}{1 + \frac{(n_0 - 1)}{N}}$$

- $\mathbf{n_0}$ is Cochran's sample size recommendation
- **N** is the population size
- **n** is the new, adjusted sample size

Yamane's Formula

$$n = \frac{N}{1 + N(e)^2}$$

Where:

- e = precision level
- N = population size.

Generalisation

- Generalization is applied by researchers in an academic setting. It can be defined as the extension of research findings and conclusions from a study conducted on a sample population to the population at large.
- Generalization is an act of reasoning that involves drawing broad conclusions from particular instances—that is, making an inference about the unobserved based on the observed. In nursing and other applied health research, generalizations are critical to the interest of applying the findings to people, situations, and times other than those in a study.

Models of Generalization

- Firestone (1993) developed a typology depicting three models of generalization that provides a useful framework for considering generalizations in quantitative and qualitative studies. There are three models
- Statistical generalization- statistical generalization relies on data to draw conclusions

- Analytical generalization- analytical generalization uses reasoning and logic
- Transferability generalization- transferability generalization applies findings to different contexts.

Multiple Choice Questions

1. **What is non-probability sampling?**

   a) Sampling technique based on random selection

   b) Sampling technique where every individual has an equal chance of being selected

   c) Sampling technique where not every individual in the population has an equal chance of being included

   d) Sampling technique based on stratification

2. **Which of the following is a type of non-probability sampling method?**

   a) Simple random sampling

   b) Stratified random sampling

   c) Convenience sampling

   d) Systematic sampling

3. **Convenience sampling is characterized by:**

   a) Random selection of participants

   b) Systematic selection of participants

   c) Choosing participants based on availability and accessibility

   d) Ensuring equal representation from all strata

4. **What is the primary disadvantage of convenience sampling?**

   a) Time-consuming

   b) Requires extensive resources

   c) May lead to biased results

   d) Provides accurate representation of the population

5. **Purposive sampling is used when:**

   a) Researchers want to ensure every individual has an equal chance of being selected

   b) Researchers want a representative sample

   c) Researchers have specific criteria for selecting participants

   d) Researchers are interested in random selection

6. **What is snowball sampling?**

   a) A sampling method where individuals are selected in a sequence

   b) A sampling method based on random allocation

c) A sampling method where existing participants recruit new participants

d) A method used to eliminate bias in sampling

7. **Quota sampling involves**

   a) Randomly selecting participants from predefined categories

   b) Ensuring equal representation from all strata

   c) Setting specific quotas for different demographic groups

   d) Systematically selecting every nth participant

8. **Which non-probability sampling method involves selecting participants based on their unique characteristics or experiences?**

   a) Quota sampling

   b) Snowball sampling

   c) Purposive sampling

   d) Convenience sampling

9. **In which situation would researchers likely use accidental sampling?**

   a) When a random sample is needed

   b) When participants are intentionally chosen

   c) When time and resources are limited

   d) When a representative sample is required

10. **Which of the following is a drawback of purposive sampling?**

    a) Lack of researcher control

    b) Potential for biased results

    c) Requires extensive resources

    d) Provides a representative sample

11. **What is a characteristic of judgmental sampling?**

    a) Random selection

    b) Deliberate selection based on expert judgment

    c) Proportional representation

    d) Systematic selection

12. **Which non-probability sampling method aims to include diverse perspectives by purposefully selecting individuals from different subgroups?**

    a) Purposive sampling

    b) Quota sampling

    c) Stratified sampling

    d) Convenience sampling

13. What is a key feature of accidental sampling?

a) Random selection

b) Deliberate selection based on specific criteria

c) Unplanned and haphazard selection

d) Systematic selection

14. In which non-probability sampling method does the researcher select participants based on their availability and willingness to participate?

a) Snowball sampling

b) Convenience sampling

c) Purposive sampling

d) Quota sampling

15. Which non-probability sampling method involves dividing the population into subgroups and then selecting participants from each subgroup based on specific criteria?

a) Convenience sampling

b) Snowball sampling

c) Stratified sampling

d) Judgmental sampling

Questions

1. What are the primary reasons for relying on available subjects in non-probability sampling methods?
2. Could you explain the types of non-probability sampling and their respective advantages and limitations in quantitative research?
3. In quantitative studies, how does sample size requirement differ from that of qualitative studies, and what factors influence these requirements?
4. What methods or techniques can researchers use to estimate an appropriate sample size for quantitative research studies?
5. Contrast the challenges in estimating sample size for qualitative studies compared to quantitative ones.
6. How do researchers address the issue of generalizability in non-probability sampling methods?
7. Explain the concept of transferability in relation to generalization in research studies.
8. Could you outline the differences between statistical generalization and analytical generalization in the context of research findings?

9. What role does the researcher's judgment play in determining the generalizability of findings in non-probability sampling?
10. How might the lack of random selection impact the generalizability of research findings in non-probability sampling?

Chapter 22

Data Collection Process and Errors

Data Collection

- ❖ Data are special types of information,generally obtained through observation, surveys, enquiries or are generated as a result of human activity.
- ❖ Collection of data and other pertinent information should be started after deciding on the objective -(G.L.RAY,1999)
- ❖ Data collection is the process of gathering and measuring information on variables of interest, in an established systematic fashion that enables one to answer stated research questions, test hypotheses, and evaluate outcomes.

Process of Data Collection

- ❖ Define Objectives and Research Questions:
  - ▷ Clearly articulate the goals of the research or the specific questions you want to answer through data collection. This helps guide the entire process.
- ❖ Review Existing Literature:
  - ▷ Conduct a thorough review of existing literature to understand what is already known in the field and identify gaps in knowledge. This informs the development of research hypotheses or objectives.
- ❖ Select a Data Collection Method:
  - ▷ Choose the most appropriate method(s) for collecting data based on the research objectives. Common methods include surveys, interviews, observations, experiments, and secondary data analysis.
- ❖ Develop Data Collection Instruments:
  - ▷ Create tools or instruments tailored to the chosen method(s). For surveys, this might involve designing questionnaires; for interviews, preparing interview guides; and for observations, defining observation protocols.
- ❖ Pilot Testing:
  - ▷ Before full-scale data collection, conduct a pilot test with a small sample to identify and address any issues with the instruments, such as confusing questions or ambiguous instructions.

- Selection and Training of Data Collectors:
  - If applicable, select and train individuals (data collectors) responsible for gathering information. Ensure they understand the research objectives, are familiar with the instruments, and adhere to ethical standards.
- Sampling:
  - Define the target population and determine the sample size and sampling technique. Random sampling, stratified sampling, or other methods may be used based on the research design.
- Data Collection:
  - Implement the data collection plan by administering surveys, conducting interviews, making observations, or employing other chosen methods. Collect data in a consistent and systematic manner to maintain reliability.
- Quality Control:
  - Implement measures to ensure data quality, such as double-checking entries, monitoring data collectors, and validating responses. This step helps minimize errors during the collection process.
- Data Entry and Management:
  - Enter collected data into a database or analysis software. Establish a data management system to organize, store, and safeguard the data.
- Data Cleaning:
  - Examine the dataset for errors, inconsistencies, or outliers. Clean the data by correcting or removing any inaccuracies.
- Data Analysis:
  - Conduct statistical or qualitative analysis, depending on the nature of the data. Use appropriate tools and techniques to derive meaningful insights from the collected information.
- Interpretation and Reporting:
  - Interpret the results in the context of the research objectives. Prepare a report or presentation summarizing key findings, conclusions, and implications.
- Dissemination:
  - Share the findings with relevant stakeholders, whether through academic publications, presentations, reports, or other means, depending on the intended audience.

Field Investigators

- A field investigators responsibility is to conduct investigations on field (i.e., in a natural environment) to information about a person, place, or situation.
- Collaborating with the local administration and communicating with the Field Investigation Program for monitoring the investigation process as per the guidelines

- ❖ Involved in a qualitative method of data collection that aims to observe, interact and understand people while they are in a natural environment
- ❖ Making recommendations on additional field tasks and assigning additional staff for carrying out the investigation if needed

Good Field Investigator

1. Selection
2. Training
3. Supervision
4. Evaluation

Selection

The field investigator should,

- ❖ Develop job specification for the project, taking into the Account the mode of Data Collection
- ❖ Decide what Characteristics the field workers should have

Recruit Appropriate Individuals.

- ❖ Field investigators play a pivotal role in data collection. It is essential to select individuals with the right qualifications, skills, and understanding of the research objectives.
- ❖ Consideration should be given to factors such as education, experience, language proficiency, and cultural sensitivity.

Training

- ❖ Proper training is crucial to ensure that field investigators understand the research objectives, data collection methods, and ethical considerations.
- ❖ Training programs should cover interview techniques, survey administration, observation skills, and any specific tools or instruments being used.
- ❖ Interviewers should be Trained to Make opening remarks that will convince potential respondents that their Participation is Important
- ❖ Asking the questions like,
- ❖ Be thorougly familiar with Questionnaire
- ❖ Ask the Questions in the order aslike in the Questionnaire
- ❖ Use the Exact Wordings given in thr Questionnaire
- ❖ Read Each Questions slowly
- ❖ Repeat Questions that are not Understood
- ❖ Ask every Applicable Questions
- ❖ Follow Instructions Probe Carefully

Supervision

Supervision means that,

- Quality Control and Editing- Checking to see if the field procedures are being properly implemented.
- Sampling Control - Supervisor attempts to ensure that the interviewer strictly following the sampling plan.
- Control of Cheating - Can be Minimized by Proper Training,Supervision, and Validation.
- Central Office Control - Supervisors provide quality and Cost control information to Central office.

Supervisors

- Calls, 10-25% of the respondents to inquire whether the field workers actually conducted the interviews.
- Asks about the length and quality of the interview, reaction to the interviewer
- The demographic information is cross-checked against information reported by the interviewer on the Questionnaires.
- Continuous supervision is necessary to monitor the progress of data collection, address any issues or concerns, and ensure consistency in the application of data collection methods.
- Supervisors should be available to provide support, guidance, and clarification on research protocols.

Evaluation

- Cost and Time: The Interviewers can be compared in terms of Total cost per Completed Interview.
- Response Rates: It is important to Monitor response rates on a timely Basis so that corrective actions can be taken.
- Quality of Interviewing: Direct Observation on the Interviewing Process.
- Quality of Data: The Completed Questionnaries of each interviewer should be evaluated for Quality.

Introduction To Online Data Collection:

- Online data collection involves using internet-based tools and platforms to gather information. This can include online surveys, interviews, and observations.
- Advantages include cost-effectiveness, speed, and the ability to reach a wider audience. However, it's essential to consider issues such as digital literacy, internet access, and potential biases in online samples.

Errors and Biases During Data Collection

- Sampling Bias: Occurs when the sample selected is not representative of the population, leading to skewed results.
- Non-response Bias: Arises when certain individuals or groups are less likely to participate, leading to a biased sample.
- Measurement Errors: Stem from inaccuracies in data collection tools, questions, or the interpretation of responses.
- Observer Bias: Occurs when the presence of the observer influences the behavior of the participants.

Multiple Choice Questions

1. **What is a critical step in ensuring the accuracy and reliability of data collection?**

   a) Data analysis

   b) Data interpretation

   c) Selection of field investigators

   d) Sampling technique

   Answer: C. Selection of field investigators

2. **What is the purpose of training field investigators in the data collection process?**

   a) To interpret data

   b) To analyze data

   c) To ensure understanding of research objectives and methods

   d) To supervise other investigators

   Answer: C. To ensure understanding of research objectives and methods

3. **What role does supervision play in the data collection process?**

   a) Data analysis

   b) Quality control

   c) Monitoring progress and ensuring consistency

   d) Data entry

   Answer: C. Monitoring progress and ensuring consistency

4. **How can the performance of field investigators be assessed during data collection?**

   a) Through data analysis

   b) Through continuous training

   c) Through regular evaluation

   d) Through online surveys

   Answer: C. Through regular evaluation

5. **What is an advantage of online data collection?**

   a) Cost-effectiveness and speed

   b) Limited reach

   c) Dependence on physical location

   d) Low digital literacy requirement

   Answer: A. Cost-effectiveness and speed

6. **What potential bias should researchers be aware of in online data collection?**

   a) Observer bias

   b) Sampling bias

   c) Non-response bias

   d) Measurement errors

   Answer: C. Non-response bias

7. **Which error may arise from inaccuracies in data collection tools or questions?**

   a) Sampling bias

   b) Non-response bias

   c) Measurement errors

   d) Observer bias

   Answer: C. Measurement errors

8. **What is the primary purpose of pilot testing in the data collection process?**

   a) Full-scale data collection

   b) Identifying and addressing issues with data collection instruments

   c) Data analysis

   d) Data interpretation

   Answer: B. Identifying and addressing issues with data collection instruments

9. **What does the process of sampling involve?**

   a) Data entry

   b) Selecting field investigators

   c) Determining the target population and sample size

   d) Training data collectors

   Answer: C. Determining the target population and sample size

10. **Which step involves entering collected data into a database or analysis software?**
    a) Data cleaning
    b) Data entry and management
    c) Quality control
    d) Data analysis
    Answer: B. Data entry and management

11. **What step involves examining the dataset for errors, inconsistencies, or outliers?**
    a) Data entry and management
    b) Data cleaning
    c) Quality control
    d) Sampling
    Answer: B. Data cleaning

12. **What is the purpose of data analysis in the data collection process?**
    a) Data interpretation
    b) Identifying and addressing issues with data collection instruments
    c) Deriving meaningful insights from the collected information
    d) Monitoring progress and ensuring consistency
    Answer: C. Deriving meaningful insights from the collected information

13. **How can errors and biases be minimized during data collection?**
    a) Increase observer bias
    b) Implement rigorous training and quality control measures
    c) Ignore sampling techniques
    d) Exclude pilot testing
    Answer: B. Implement rigorous training and quality control measures

14. **What is the last step in the data collection process before dissemination of findings?**
    a) Data cleaning
    b) Data analysis
    c) Data entry and management
    d) Interpretation and reporting
    Answer: D. Interpretation and reporting

15. Why is adherence to ethical principles crucial during data collection?

a) To increase sampling bias

b) To improve data cleaning

c) To safeguard the confidentiality of participants

d) To ignore potential biases

Answer: C. To safeguard the confidentiality of participants

Short Notes

1. What is the importance of selecting qualified field investigators in the data collection process?
2. Why is training essential for field investigators, and what key skills should they acquire?
3. How does supervision play a role in ensuring the quality of collected data during field investigations?
4. What criteria should be considered when evaluating the performance of field investigators in data collection?
5. What are the advantages of online data collection methods over traditional approaches?
6. What challenges might arise in online data collection, and how can they be addressed?
7. How can researchers minimize errors during the data collection process?
8. What is the significance of addressing biases in data collection, and how can it be achieved?
9. Can you provide examples of common biases that may occur during field investigations?
10. 10)In what ways can technology be utilized to improve the accuracy and efficiency of data collection?

Chapter-23

Reliability and Validity of Measures

The concept of "Testing goodness of measures through item analysis - Reliability and Validity" is a fundamental aspect of research methodology, particularly in the fields of psychology, education, and social sciences. This concept revolves around the evaluation and improvement of measurement instruments, such as tests, questionnaires, or surveys, to ensure they accurately and consistently measure what they are intended to measure. Let's break down the main components of this concept:

Introduction to Item Analysis, Reliability, and Validity

Item Analysis

Purpose: Item analysis involves examining each question (item) in a measurement instrument to assess its effectiveness and contribution to the overall construct being measured.

Methods: Techniques such as item difficulty, item discrimination, and item-total correlations are used to evaluate items.

Outcome: The goal is to retain items that are effective and eliminate or revise those that are not, thereby enhancing the overall quality of the measurement tool.

Reliability

Definition: Reliability refers to the consistency and dependability of a measurement instrument. It indicates how reliably the tool can reproduce similar results under consistent conditions.

Types

Test-Retest Reliability: Measures stability over time.

Inter-Rater Reliability: Assesses the extent of agreement among different observers.

Internal Consistency: Evaluates how well the items on a test measure the same construct (e.g., Cronbach's alpha).

Importance: A reliable instrument yields consistent results, which is crucial for the credibility and reproducibility of research findings.

Validity

Definition: Validity is about the accuracy of a measurement tool, i.e., whether it measures what it is supposed to measure.

Types

- Content validity
- Construct validity
  - Convergent validity
  - Discriminant validity
- Criterion Validity
  - Concurrent validity
  - Predictive validity
- Face validity
- External validity
  - Population validity
  - Ecological validity
  - Temporal validity
- Internal validity
- Factorial validity
- Nomological validity

Importance: Validity is essential for the relevance and accuracy of the conclusions drawn from the research.

Why are they Important?

Both reliability and validity are crucial in research methodology for several reasons:

Quality of Research: They determine the quality and trustworthiness of the data collected.

Interpretation of Results: High reliability and validity are necessary for the accurate interpretation and generalization of research findings.

Informed Decisions: In applied settings, such as education and clinical psychology, these measurements inform critical decisions and interventions.

Validity

Validity in research methodology refers to the extent to which a test or instrument measures what it is intended to measure. It is a critical factor in the development and evaluation of any measurement tool, such as tests, surveys, or questionnaires, especially in fields like psychology, education, and social sciences. Here's a detailed look at the different types of validity and their importance:

Types of Validity

1. Content Validity

Definition: Content validity refers to the extent to which a measurement tool covers the entire range of the concept it's supposed to measure.

Evaluation: It involves subject matter experts assessing whether the test items represent all facets of the given construct.

Example: In a math test, content validity would mean the test includes a balanced representation of all topics relevant to the curriculum being tested.

2. Construct Validity

Construct validity is a crucial concept in research methodology, particularly in the fields of psychology, education, and the social sciences. It refers to the degree to which a test or measurement tool accurately measures the theoretical construct or psychological concept it is intended to measure. This type of validity is essential for ensuring that the inferences made based on the test scores are meaningful and accurate. Construct validity encompasses several aspects, each of which contributes to understanding and validating the measurement tool.

Definition: Construct validity assesses whether a tool truly measures the theoretical construct it's designed to measure.

Sub-types

Convergent Validity:

Definition: Convergent validity assesses whether a test correlates with other measures that it is theoretically expected to correlate with.

Example: A new anxiety scale should have a high correlation with existing validated anxiety scales.

Discriminant Validity (or Divergent Validity):

Definition: Discriminant validity tests whether concepts or measurements that are not supposed to be related are actually unrelated.

Example: The aforementioned anxiety scale should not correlate strongly with unrelated constructs, such as a measure of physical health.

Importance of Construct Validity

Theoretical Alignment: Construct validity ensures that the measurement tool aligns with the theoretical understanding of the construct.

Accurate Measurement: It is critical for ensuring that the tool measures what it is intended to measure, not some other, unrelated construct.

Research Credibility: High construct validity contributes to the credibility and reliability of research findings.

Basis for Inferences: It provides a solid foundation for making psychological and educational inferences based on the data collected from the measurement tool.

3. Criterion-related Validity

Criterion-related validity is a key type of validity in research methodology, particularly when the goal is to predict outcomes or when a new measurement tool is evaluated against an established standard. This form of validity assesses how well one measure predicts an outcome (or correlates with a criterion) that is already known.

Definition: This type of validity examines the effectiveness of a measure in predicting or correlating with a known criterion.

There are two main types of criterion-related validity: concurrent validity and predictive validity.

Concurrent Validity

Definition: Concurrent validity refers to the extent to which a test's results correlate with those of an established measure when both are administered at roughly the same time.

Application: This type is often used when developing a new measure that aims to be more efficient or accessible than an existing one.

Example: If a new brief depression inventory correlates highly with scores from a well-established depression scale when administered to the same group at the same time, it has high concurrent validity.

Predictive Validity

Definition: Predictive validity measures how well a test predicts future outcomes. It's about the test's ability to effectively forecast a criterion that will occur in the future.

Application: This type is crucial in situations where the purpose of the test is to make predictions about future behaviors or outcomes, such as in educational admissions or employment settings.

Example: The SAT exam's predictive validity is determined by how well SAT scores predict college academic performance in subsequent years.

Importance of Criterion-related Validity

Practical Decision-Making: It's essential for tests used in making practical decisions, like hiring or academic admissions, where the stakes are high, and accurate predictions are vital.

Test Development and Evaluation: Criterion-related validity is crucial for developing new tests and refining existing ones to ensure they serve their intended purpose.

Ensuring Relevance: It helps in ensuring that a test remains relevant and useful in real-world contexts by demonstrating its practical utility.

4. Face Validity

Face validity is a basic and somewhat subjective form of validity, particularly in the development and assessment of tests, surveys, and other measurement instruments. Face validity is the most basic form of validity, where the measure appears to be assessing the intended construct just by its content.

Definition: It refers to the extent to which a test appears to measure what it is supposed to measure, based solely on its content and the way it looks at face value, rather than on any empirical testing.

Note: It is more about the appearance of validity rather than being a rigorous assessment.

Example: A depression questionnaire that includes items clearly related to mood, sadness, and loss of interest has high face validity.

5. External Validity

External validity is a key concept in research methodology, especially in experimental design. It refers to the extent to which the results of a study can be generalized or extended to contexts outside of the experimental conditions. This includes different settings, populations, times, and other variables than those used in the original study. In simpler terms, external validity addresses the question: "Can the findings of this study be applied to the real world?"

Definition: External validity refers to the extent to which the results of a study can be generalized to other settings, populations, and times.

Example: If the findings of a clinical trial on a new drug are applicable to the general population beyond the trial's participants, the study has high external validity.

Types of External Validity

a. Population Validity: Concerns whether the study results can be generalized to a broader population beyond the sample used.

b. Ecological Validity: Relates to the extent to which findings can be applied to real-world settings.

It's about the naturalness of the environment in which the data was collected and whether this environment matches the one to which the results are to be applied.

c. Temporal Validity: Refers to the generalizability of the findings across different times.

It is particularly important in studies where the time factor could influence the results (e.g., studies on trends or attitudes that might change over time).

Importance of External Validity

Generalization: The main importance of external validity lies in the ability to generalize the findings of a study to broader contexts. This is crucial for the practical application of research results.

Practical Impact: High external validity ensures that the research can have real-world implications and can inform policies, educational practices, clinical interventions, etc.

Replicability: Studies with high external validity are more likely to be replicated in different settings, which is a cornerstone of scientific research.

Challenges in Achieving External Validity

Control vs. Generalizability: Often, there's a trade-off between the control over variables (high in laboratory settings) and the generalizability of the results. Highly controlled environments might not accurately reflect real-world settings.

Representative Samples: Achieving a sample that truly represents the broader population can be challenging.

Changing Times and Contexts: What is valid at one time or in one context might not hold true in another.

6. Internal Validity

Internal validity refers to the extent to which a study can establish a causal relationship between variables and demonstrate that the outcome is indeed due to the manipulation of the independent variable and not due to other factors. Essentially, internal validity is about the soundness of the study design and execution, ensuring that the conclusions drawn about cause and effect are legitimate.

Definition: Internal validity is concerned with the extent to which a causal conclusion based on a study is warranted, which is particularly relevant in experimental research.

Example: A study testing a new teaching method would have high internal validity if it can confidently attribute changes in students' performance to the teaching method, rather than other factors.

Importance of Internal Validity

Causal Inferences: High internal validity is essential for confidently asserting causal relationships in research.

Foundation for External Validity: While distinct, internal validity is a prerequisite for external validity; only a study with strong internal validity can meaningfully be generalized to other contexts.

Research Credibility: Studies with high internal validity are more likely to be viewed as credible and contribute significantly to the field.

Threats to Internal Validity

Selection Bias: Differences between groups that are not related to the manipulation of the independent variable can affect the results.

History Effects: External events occurring during the study can influence the results.

Maturation: Changes in participants over time, not related to the experimental manipulation, can affect outcomes.

Testing Effects: The act of testing or measuring participants can influence their subsequent behavior.

Instrumentation Changes: Alterations in measurement tools or techniques during the study can lead to inconsistencies.

Experimental Attrition: Loss of participants during the study can lead to biased results.

Experimenter Bias: The researcher's expectations or influences can unintentionally affect participants' performance.

7. Factorial Validity

Definition: Factorial validity is a type of construct validity. It refers to the extent to which a test measures the number of constructs it claims to measure, and the degree to which each item relates to its intended construct. This is typically assessed using factor analysis.

Evaluation Method: Factorial validity is often evaluated using exploratory or confirmatory factor analysis. These statistical methods identify the underlying structure of the data and check if items group together in ways consistent with the proposed constructs.

Example: In a personality test claiming to measure five distinct traits (like the Big Five personality traits), factorial validity would be demonstrated if factor analysis indeed shows five distinct factors, each corresponding to one of the traits.

8. Nomological Validity

Definition: Nomological validity, another aspect of construct validity, refers to the extent to which predictions from a theoretical network (or nomological net) involving the construct are confirmed. It's about the test's ability to fit into the existing network of theories and constructs.

Evaluation Method: This involves testing whether relationships predicted by theoretical models are empirically observed. It often requires a series of studies, examining the correlations of the measure with other measures and outcomes as predicted by theory.

Example: If a new measure of self-efficacy is developed, its nomological validity would be supported if it correlates positively with measures of motivation and performance, as predicted by theoretical models of self-efficacy.

Importance of Validity

The importance of validity in research cannot be overstated, as it is fundamental to the integrity and applicability of the results. Validity determines whether a research instrument, such as a test, survey, or other measurement tool, accurately measures what it is intended to measure. Here are key reasons why validity is crucial:

Accurate Measurement of Constructs: The primary purpose of most research tools is to measure abstract constructs like intelligence, satisfaction, depression, or learning outcomes. Validity ensures that these constructs are measured accurately and meaningfully.

Credibility of Research Findings: Valid research instruments lend credibility to the study's findings. Without validity, the results of a study can be misleading, misinterpreted, or invalid, leading to incorrect conclusions and potentially harmful applications.

Informed Decision-Making: In fields such as education, psychology, healthcare, and social policy, decisions are often based on data from research studies. Validity ensures that these decisions are informed by accurate and relevant data.

Generalizability of Results: High external validity allows researchers to generalize their findings to broader populations or different contexts, enhancing the applicability and impact of the research.

Theoretical Advancement: Validity is essential for theoretical development in various fields. By accurately measuring constructs, researchers can test theoretical models and hypotheses, leading to advancements in understanding.

Ethical Research Practice: Ensuring the validity of research tools is also an ethical imperative. It respects the time and effort of participants and avoids misusing resources on flawed research.

Foundation for Future Research: Studies with high validity provide a solid foundation for future research. They offer reliable data that other researchers can build upon, facilitating cumulative knowledge development.

Policy and Practice Implementation: Research with high validity is crucial for informing policy and practice. In areas like public health, education, and social policy, valid research guides effective interventions and strategies.

Resource Allocation: Valid research helps ensure that resources, such as funding and effort, are allocated effectively and not wasted on pursuing false leads or ineffective interventions.

Ensuring Validity

Ensuring validity in research is crucial for the reliability and credibility of the study's findings. Validity refers to how well a test or tool measures what it is supposed to measure. There are several strategies and steps that researchers can take to ensure the validity of their studies:

- Define the Construct Clearly: Start by clearly defining the construct or phenomenon being measured. This clarity is essential for developing an appropriate measurement tool.

Use Established Measures

- Proven Tools: Whenever possible, use measurement tools that have already been validated in previous research.
- Adaptation with Caution: If adapting existing measures, do so carefully to maintain their validity.

Employ Rigorous Test Design

- Content Validity: Involve subject matter experts to ensure that the test covers all relevant aspects of the construct.
- Pre-Testing: Conduct pilot testing or preliminary studies to refine the instrument.

Check for Reliability

- Consistency: Ensure that the measurement tool yields consistent results over time (test-retest reliability) and across different observers (inter-rater reliability).
- Internal Consistency: For tests with multiple items, check internal consistency (e.g., using Cronbach's alpha).

Validate the Instrument

- Empirical Validation: Use statistical methods to validate the instrument. This can include factor analysis for construct validity and correlational studies for criterion-related validity.
- Triangulation: Use multiple methods and sources of data to validate the findings.

Consider the Sample

- Representative Sampling: Ensure that the sample is representative of the population for whom the conclusions will be drawn.
- Adequate Sample Size: Use an appropriately large sample size to enhance the validity of the conclusions.

Address Potential Biases

- Awareness of Bias: Be aware of and address potential biases in the data collection and analysis processes.
- Blinding and Randomization: In experimental designs, use blinding and randomization to minimize biases.

Evaluate External Validity

- Generalizability: Assess the extent to which the findings can be generalized to other settings, populations, or times.

- Replication: Encourage or conduct replication studies to see if the findings hold in different contexts.
- Continuous Re-evaluation - Ongoing Process: Understand that validity is not a one-time verification. Continuously reevaluate the validity of the measurement tool, especially as new research emerges.

Documentation and Transparency

- Detailed Reporting: Report the processes of ensuring validity in detail, allowing others to assess the credibility of the findings.
- Openness to Critique: Be open to feedback and critiques, which can further strengthen the validity of the research.

Ensuring validity is a comprehensive process that involves careful design, implementation, and evaluation of research methods. It's crucial for the trustworthiness of research findings and for the advancement of knowledge in any field.

Conclusion

In conclusion, the process of testing the goodness of measures through item analysis, reliability, and validity is integral to the development of effective and reliable research instruments. It ensures that the tools used in research are both consistent and accurate in measuring the intended constructs, leading to more reliable and valid research outcomes.

Multiple Choice Questions

1. What is the primary purpose of item analysis in research methodology?

a) To determine the statistical significance of the results

b) To examine each question in a measurement instrument for effectiveness

c) To calculate the overall score of the test

d) To compare different measurement instruments

2. What does reliability refer to in the context of a measurement instrument?

a) Accuracy of the tool

b) Consistency and dependability of the tool

c) Complexity of the tool

d) Length of the tool

3. Which of the following is a type of reliability?

a) Convergent Validity

b) Test-Retest Reliability

c) Content Validity

d) Criterion-Related Validity

4. Internal consistency in a test is evaluated using which statistic?

a) Pearson's correlation coefficient

b) Cronbach's alpha

c) ANOVA

d) Chi-square test

5. Content validity is concerned with:

a) The extent to which a test correlates with other measures

b) How well the test covers the entire range of the concept it's supposed to measure

c) The appearance of the test

d) The prediction of future outcomes

6. Which type of validity involves assessing whether a test correlates with other measures that it is theoretically expected to correlate with?

a) Discriminant Validity

b) Convergent Validity

c) Internal Validity

d) Face Validity

7. Which type of validity is concerned with how well a test predicts future outcomes?

a) Concurrent Validity

b) Predictive Validity

c) Content Validity

d) Construct Validity

8. Face validity refers to:

a) The extent to which a test appears to measure what it is supposed to measure

b) The empirical testing of a test

c) The statistical analysis of a test

d) The theoretical framework of a test

9. External validity is about:

a) The internal consistency of a test

b) The generalizability of study results to other settings or populations

c) The accuracy of a test in measuring a specific construct

d) The prediction of future outcomes based on test results

10. **Internal validity is important for asserting:**

    a) Causal relationships in research

    b) Generalizability of results

    c) The consistency of test results

    d) The relevance of the test to the concept measured

11. **Which type of validity is evaluated using factor analysis?**

    a) Face Validity

    b) Criterion-Related Validity

    c) Factorial Validity

    d) Nomological Validity

12. **Nomological validity is related to:**

    a) The test's ability to fit into an existing network of theories and constructs

    b) The extent to which a test measures a single construct

    c) The relevance of test content to the measured construct

    d) The prediction of future outcomes using test results

13. **Inter-rater reliability is essential for:**

    a) Determining the stability of test results over time

    b) Ensuring agreement among different observers

    c) Assessing the complexity of the test items

    d) Evaluating the predictive power of a test

14. **Criterion-related validity is divided into which two main types?**

    a) Internal and External Validity

    b) Concurrent and Predictive Validity

    c) Content and Construct Validity

    d) Factorial and Nomological Validity

15. **What does discriminant validity test?**

    a) The relationship between related constructs

    b) The relationship between unrelated constructs

    c) The appearance of the test to untrained observers

    d) The statistical significance of test results

Short Answer Type Questions

Q: What is item analysis in research?

A: Item analysis is the process of examining each question in a measurement instrument to assess its effectiveness and contribution to the overall construct being measured.

Q: Define reliability in the context of measurement instruments.

A: Reliability refers to the consistency and dependability of a measurement instrument, indicating how reliably the tool can reproduce similar results under consistent conditions.

Q: What is content validity?

A: Content validity refers to the extent to which a measurement tool covers the entire range of the concept it is supposed to measure.

Q: What does construct validity assess?

A: Construct validity assesses whether a measurement tool truly measures the theoretical construct it is designed to measure.

Q: What is the difference between concurrent and predictive validity?

A: Concurrent validity refers to the extent to which a test's results correlate with those of an established measure when both are administered at roughly the same time, while predictive validity measures how well a test predicts future outcomes.

Q: Why is internal validity important in research?

A: Internal validity is important because it determines whether the outcomes of a study can legitimately be attributed to the manipulation of the independent variable, rather than other factors.

Q: What does factorial validity involve?

A: Factorial validity involves using factor analysis to assess whether a test measures the number of constructs it claims to measure and the degree to which each item relates to its intended construct.

Q: Why is external validity important in the generalization of research findings?

A: External validity is crucial because it determines the extent to which the findings of a study can be generalized beyond the specific conditions of the study. High external validity means that the results can be applied to other settings, populations, and times, enhancing the broader applicability and impact of the research.

Q: Explain the significance of test-retest reliability in research instruments.

A: Test-retest reliability measures the stability of a research instrument over time. It involves administering the same test to the same group of people at two different points in time and then correlating the scores. This type of reliability is crucial for assessing whether the instrument can produce consistent results over time.

Chapter-24

Types of Reliability and Validity Enhancement

Types of Reliability

There are four common methods for estimating the reliability coefficient of test scores

- Test – retest reliability
- Internal consistency reliability
- Parallel – forms reliability or Alternate – forms reliability or Equivalent – forms reliability or Comparable – forms reliability
- Scorer reliability

Test – Retest Reliability

- In Test- retest reliability the single form of the test is administered twice on the same sample with a reasonable time gap.
- Two administrations of the same test yield two independent sets of scores.
- The two sets when correlated give the value of the reliability coefficient
- The reliability coefficient is also known as the temporal stability coefficient
- Time too short – increase reliability coefficient
- Time too long – Lower the reliability coefficient
- Convenient time gap between two administrations – Fortnight

Internal Consistency Reliability

- It indicates the homogeneity of the test
- If all the items of the test measure the same function or the trait, the test is said to be a homogeneous one and its internal consistency reliability would be pretty high.

The common methods for estimating the internal consistency reliability are

- Split- half method
- Rulon and Flanagan Formula

- Kuder – Richardson Formula
- Coefficient alpha

1. Split- Half Method

The most common method of estimating the internal consistency reliability is the split- half method

- The common way of splitting the test is odd- even method
- In this method all the odd numbered items constitute one part of the test and all the even number items constitute the other part of the test.
- Each examinee thus receives two scores
- Product moment correlation is computed to obtain the reliability of the half test.
- On the basis of this half test reliability, the reliability of the whole test is estimated

The Spearman – Brown prophecy formula is used for estimating the reliability of the whole test

$$\text{Reliability of the whole test} = \frac{2 \times \text{reliability of half test}}{1 + \text{reliability of half test}}$$

2. Rulon and Flanagan Formula

- Both these formulas provide the reliability of the whole test and not the half test and both formulas estimate the reliability coefficient on the basis of proportion of error variance in total variance of the test.
- The lesser the error variance, the higher the true variance and therefore, the higher will be the reliability

3. Kuder- Richardson Formula

Their formulas 20 and 21 are very popular

K – R 20 is the basic formula and K- R 21 is the modified formula of k- R 20

The main requirements for the use of K R formula are

- All items of the test should be homogeneous
- Items should be scored as +1 or 0, that is, all correct answered should be scored as +1 and all the incorrect answers should be scored as zero
- Items should not be very much in their indices of difficulty

4. Co-Efficient Alpha

The Kuder- Richardson formulas are applicable to the tests whose items are scored as 0 or 1 (right or wrong)

- Some tests however have multiple- score items (Sometimes, Usually, Rarely, Never)
- For calculating the reliability of such tests, a generalized formula known as Coefficient alpha has been formulated (Cronbach1951 , Kaiser & Michael 1975)

Alternate - Forms Reliability

Alternate – Forms reliability is known by various names such as

1. Parallel – forms reliability
2. Equivalent – forms reliability
3. Comparable - forms reliability

- It requires that the test be developed in two forms, which should be comparable or equivalent
- Two forms of the tests are administered to the same sample either immediately on the same day or with a time interval of usually a fortnight
- When the reliability is calculated on the basis of the data collected immediately it is called alternate – form (immediate) reliability and if it is calculated on the basis of the data collected after a gap of fortnight it is called alternate – form (delayed) reliability.
- The biggest problem in Alternate – Forms tests is how to make both the forms equivalent in true sense
- Gulliksen (1950) has defined parallel tests as tests having equal means, equal variances and equal inter- item correlations.
- Freeman (1962) has listed the following criteria for judging whether or not the two forms of the test are parallel.
- The number of items in both the forms should be the same
- Items in both the forms should have uniformity regarding the content, the range of difficulty and the adequacy of sampling.
- Distribution of the indexes of difficulty of items in both should be similar.
- Items in both the forms should be of equal degree of homogeneity, which can be shown either by inter – item correlation or by correlating each item with sub test scores or with total test scores.
- Means and standard deviations of both the forms should be equal or nearly so.
- Mode of administration and scoring of both the forms should be uniform.

Factors Affecting the Validity and Reliability of Research Instruments

1. **Instrument Design**: The clarity, appropriateness, and precision of the research tool impact its reliability and validity.

2. **Sampling**: The representativeness of the sample chosen for the study affects the generalizability of results.
3. **Data Collection Procedures**: Consistency in applying data collection methods ensures reliability, while the appropriateness of methods contributes to validity.
4. **Participant Response Bias**: The tendency of participants to respond in a socially desirable manner can impact the accuracy of data.
5. **Instrument Calibration**: Regular checks and adjustments to ensure instruments measure consistently over time enhance reliability.
6. **Scoring Consistency**: Clear and standardized scoring procedures promote reliability in the interpretation of results.
7. **Researcher Bias**: The researcher's own perspectives and expectations can influence the interpretation of data, affecting both validity and reliability.
8. **External Factors**: Environmental conditions, participant mood, or unforeseen events may introduce variability, impacting the reliability of results.

Considering and addressing these factors is crucial for maintaining the credibility and robustness of research findings.

Strategies for Enhancing Validity and Reliability of Measures

1. **Pilot Testing**: Conduct a preliminary test of the research instrument to identify and address any issues before full-scale data collection.
2. **Clear Operational Definitions**: Clearly define concepts and variables to ensure consistent interpretation by both researchers and participants.
3. **Randomization:** Randomly assign participants to conditions to control for confounding variables and enhance the reliability of experimental designs.
4. **Standardized Procedures**: Implement consistent and standardized procedures for data collection to improve the reliability of measurements.
5. **raining and Calibration:** Ensure that researchers and data collectors are well-trained and calibrated to minimize variations in data collection practices.
6. **Use of Established Instruments:** Utilize validated and reliable measurement tools that have undergone rigorous testing in previous research.
7. **Multiple Measures**: Employ multiple measures for each construct to enhance the validity of findings through converging evidence.

Validity and Reliability in Qualitative Research

1. **Triangulation**: Use multiple data sources, methods, or researchers to cross-verify findings and enhance the trustworthiness of qualitative data.
2. **Member Checking**: Share research findings with participants to validate interpretations and ensure the accuracy of qualitative data.

3. **Rich Descriptions**: Provide detailed and contextually rich descriptions of the study setting, participants, and processes to enhance the validity of qualitative findings.
4. **Reflexivity**: Acknowledge and consider the researcher's influence on the study, promoting transparency and credibility in qualitative research.
5. **Peer Review**: Seek feedback from peers or experts in the field to ensure the rigor, credibility, and validity of qualitative research interpretations.
6. **Saturation:** Continue data collection until saturation is reached, ensuring comprehensive exploration of the research topic and enhancing the reliability of qualitative findings.

Applying these strategies can contribute to the robustness and credibility of both quantitative and qualitative research.

Multiple Choice Questions:

1. **What type of reliability assesses the consistency of results across different items within the same test?**

   a) Internal reliability

   b) External reliability

   c) Test-retest reliability

   d) Inter-rater reliability

   Answer: a) Internal reliability

2. **Which type of reliability evaluates the stability of measurements over time, usually through repeated assessments?**

   a) Concurrent reliability

   b) Split-half reliability

   c) Test-retest reliability

   d) Alternate forms reliability

   Answer: c) Test-retest reliability

3. **In a research study, if two different researchers independently measure the same set of behaviors and produce consistent results, it is an example of:**

   a) Internal reliability

   b) External reliability

   c) Inter-rater reliability

   d) Intraclass reliability

   Answer: c) Inter-rater reliability

4. **Splitting a test into two equivalent halves and checking for consistency in scores between the halves measures:**

   a) Internal reliability

   b) Split-half reliability

   c) Test-retest reliability

   d) Face validity

   Answer: b) Split-half reliability

5. **What factor refers to the extent to which a research instrument accurately measures the intended concept or variable?**

   a) Reliability

   b) Validity

   c) Generalizability

   d) Replicability

   Answer: b) Validity

6. **Participant response bias can negatively impact:**

   a) Reliability

   b) Validity

   c) Generalizability

   d) Consistency

   Answer: b) Validity

7. **In research, external factors such as environmental conditions and participant mood may affect the:**

   a) Reliability

   b) Validity

   c) Generalizability

   d) Consistency

   Answer: a) Reliability

8. **What strategy involves conducting a preliminary test of the research instrument to identify and address potential issues?**

   a) Randomization

   b) Pilot Testing

   c) Triangulation

   d) Member Checking

   Answer: b) Pilot Testing

9. **Using established and validated measurement tools is a strategy to enhance:**

   a) Reliability

   b) Validity

   c) Generalizability

   d) Consistency

   Answer: b) Validity

10. **What qualitative research strategy involves seeking feedback from peers or experts to ensure the rigor and credibility of interpretations?**

    a) Triangulation

    b) Peer Review

    c) Saturation

    d) Reflexivity

    Answer: b) Peer Review

11. **Continuation of data collection until no new themes or information emerges is known as:**

    a) Triangulation

    b) Peer Review

    c) Saturation

    d) Reflexivity

    Answer: c) Saturation

Short Notes

1. What is validity in the context of research instruments?
2. How does participant response bias impact the quality of data in research?
3. Name one external factor that can influence the reliability of research results.
4. What does pilot testing aim to achieve in the enhancement of research instrument reliability?
5. Why is the use of established instruments considered a strategy for improving measurement validity?
6. In qualitative research, what is the purpose of seeking peer review?
7. Define saturation in the context of qualitative research data collection.

Chapter 25

Data Coding and Processing

Introduction

Within the field of research methodology, data coding is an essential step that converts raw information into a format that can be analysed. By applying codes or labels to data points, this methodical approach makes it easier to extract insightful information. In both qualitative and quantitative research, data coding is essential for classifying, organising, and interpreting the information gathered. Coding is a process that connects raw data to meaningful analysis; it is used in both quantitative and qualitative research to identify themes and assign numbers. The context for comprehending the importance of data coding in identifying trends, connections, and themes that strengthen the validity of research findings is provided by this introduction.

Key aspects of data coding

Quantitative Research:

- In quantitative studies, coding often involves assigning numerical values to responses or variables.
- For example, assigning "1" for male and "2" for female in a gender variable.

Qualitative Research:

- In qualitative research, coding involves categorizing textual or visual data into themes or patterns.
- Researchers create a coding scheme to represent recurring ideas, concepts, or behaviors.
- Open Coding vs. Axial Coding vs. Selective Coding:
- In qualitative research, coding may progress through stages like open coding (initial exploration), axial coding (finding connections), and selective coding (focusing on core themes).

Coding Sheets or Software:

- Researchers may use coding sheets or specialized software (e.g., NVivo, ATLAS.ti) to manage and organize codes efficiently.

- **Inter-Coder Reliability:**
- If multiple researchers are involved, establishing inter-coder reliability ensures consistency in coding decisions.

Coding Categories:

- Codes may represent different aspects, such as demographics, behaviors, attitudes, or themes emerging from qualitative data.

Data Cleaning:

Cleaning the coded data involves reviewing and correcting errors or inconsistencies in the coding process.

Statistical Analysis:

Once coded, quantitative data can be subjected to statistical analysis, enabling researchers to identify patterns, correlations, or trends.

Thematic Analysis:

In qualitative research, thematic analysis involves exploring, analyzing, and reporting patterns within the data, often derived from the coded themes.

Continuous Iteration:

Coding is not a one-time process; it may be an iterative one where researchers refine codes as they gain a deeper understanding of the data.

Steps in data coding:

The process of data coding in research methodology typically involves several key steps:

1. Data Preparation:

Ensure that the collected data is well-organized and ready for coding. This may involve cleaning and structuring raw data.

2. Define Coding Categories:

Clearly define the coding categories based on the research objectives. These categories may include variables, themes, or concepts.

3. Coding Scheme Development:

Create a coding scheme that outlines the rules and guidelines for assigning codes. This document provides a reference for consistency during the coding process.

4. Pilot Coding:

Conduct a pilot coding phase on a small subset of the data. This helps identify potential issues, refine the coding scheme, and ensure inter-coder reliability.

5. Coding Process:

Apply the coding scheme systematically to the entire dataset. This can be done manually or using specialized software.

6. Inter-Coder Reliability:

If multiple coders are involved, assess inter-coder reliability to ensure consistency in coding decisions. Resolve any discrepancies through discussions or additional training.

7. Data Verification:

After coding, verify the accuracy of the coded data. Check for errors, inconsistencies, or missing information.

8. Documentation:

Maintain detailed documentation of the coding process, including the coding scheme, any modifications made during coding, and decisions related to ambiguous data points.

9. Data Analysis:

Once coding is complete, proceed to analyze the coded data. This may involve statistical analysis for quantitative data or thematic analysis for qualitative data.

10. Refinement and Iteration:

Iteratively refine codes as needed, especially if new insights emerge during the analysis. Coding is often an evolving process that adapts to the evolving understanding of the data.

11. Report Findings:

Incorporate the coded data into research reports, ensuring transparency about the coding process and presenting findings in a clear and comprehensible manner.

Data exploration

Introduction:

Data exploration is an essential first step in research methodology, similar to setting out on an exploration of discovery within a dataset. This procedure involves searching over and understanding the distributions, patterns, and traits of the gathered data. Researchers can obtain important insights that set the way for well-informed analysis and interpretation by exploring the particulars of the dataset. The importance of data exploration in revealing hidden trends, outliers, and relationships which in turn shape the course of the research project is highlighted in this introduction.

Key aspects and techniques involved in data exploration

Data Summarization:

Begin by summarizing key statistics such as mean, median, mode, range, and standard deviation. This provides a high-level overview of the central tendencies and variability in the data.

Univariate Analysis:

Analyze individual variables in isolation. This includes exploring the distribution of categorical variables through frequency tables and visualizing numerical variables using histograms or box plots.

Bivariate Analysis:

Explore relationships between pairs of variables. Scatter plots, correlation coefficients, and cross-tabulations help identify potential associations or patterns.

Multivariate Analysis:

Examine interactions among three or more variables. Techniques like regression analysis or factor analysis can reveal complex relationships within the dataset.

Outlier Detection:

Identify and investigate outliers, data points that significantly deviate from the overall pattern. Outliers can provide insights into data quality issues or highlight unique cases.

Data Distribution:

Understand the distribution of data, whether it follows a normal distribution or exhibits skewness. This knowledge is crucial for choosing appropriate statistical tests.

Visualization Techniques:

Utilize various visualization tools such as scatter plots, bar charts, heatmaps, and histograms to represent the data graphically. Visualization aids in recognizing patterns and trends.

Missing Data Analysis:

Assess the extent of missing data and understand its patterns. Missing data can impact the reliability of analyses, and appropriate strategies for handling it may need to be implemented.

Data Transformation:

Apply transformations if needed, such as logarithmic transformations or normalization, to address issues like skewed distributions.

Data Quality Check:

Scrutinize the dataset for errors, inconsistencies, or anomalies. Cleaning the data ensures the accuracy and reliability of subsequent analyses.

Data Mining Techniques:

Consider using data mining techniques to uncover hidden patterns or trends that might not be immediately apparent. Clustering or association rule mining are examples of such techniques.

Documentation:

Document all steps taken during data exploration, including the rationale behind decisions. This documentation aids in transparency and allows for reproducibility of analyses.

Data editing:

Data editing involves reviewing, cleaning, and organizing collected data to ensure its accuracy, consistency, and completeness. The primary goal of data editing is to identify and rectify errors, inconsistencies, and missing information before analysis.

The key aspects of data editing are:

Identification of Errors:

- ❖ Data editing begins with the identification of errors or discrepancies in the dataset. Errors can include typos, outliers, missing values, and inconsistencies.
- ❖ Researchers carefully review each data point to identify anomalies or discrepancies that may affect the quality of the data.

Types of Errors:

- ❖ Errors can be classified as logical errors (inconsistencies in responses), transcription errors (typos or data entry mistakes), and factual errors (incorrect information provided by respondents).
- ❖ Researchers categorize errors to determine appropriate correction methods.

Coding and Categorization:

- ❖ If the data involves categorical variables, researchers ensure that the coding is consistent and accurate. This includes checking for proper labeling and assigning numerical codes appropriately.
- ❖ Reviewing codebooks and data dictionaries helps maintain consistency in coding across the data set.

Outlier Detection:

- ❖ Outliers are extreme values that deviate significantly from the majority of the data. Identifying outliers is crucial as they can disproportionately impact statistical analyses.

- Statistical techniques, such as visual inspection, box plots, or z-scores, may be employed to detect and address outliers appropriately.

Missing Data Handling:

- Missing data can occur for various reasons, such as non-response or errors in data collection. Researchers need to decide how to handle missing values.
- Options for dealing with missing data include imputation (replacing missing values with estimated values), deletion of incomplete cases, or conducting analyses that account for missing data.

Consistency Checks:

- Data editing involves checking for consistency within the dataset. This includes ensuring that variables are logically related and that responses are coherent.
- Running consistency checks involves examining relationships between variables and making corrections where inconsistencies are found.

Data Transformation:

- Sometimes, data may need to be transformed to meet the assumptions of statistical analyses or to create new variables for analysis.
- Transformation may include converting units of measurement, normalizing data, or creating derived variables.

Data Validation:

- Validation involves confirming that the data conforms to specific criteria or rules set during the data collection phase.
- Researchers compare data values against predefined rules or ranges to validate its accuracy and adherence to study requirements.

Documentation:

- Comprehensive documentation of the data editing process is essential for transparency and reproducibility.
- Researchers maintain detailed records of the steps taken during data editing, including any corrections or transformations made to the dataset.

Quality Control:

- Implementing quality control measures ensures that the data editing process is consistent and that multiple individuals review and validate the edited dataset.
- Independent reviewers or a data editing team can help ensure the quality and reliability of the final dataset.

Methods of Data processing in Quantitative studies

Effective data processing is essential for producing reliable and valid results in quantitative studies. It involves a systematic approach to handling data, from initial entry to final interpretation, and often requires a combination of manual and automated methods to ensure accuracy and efficiency.

Data processing in quantitative studies involves organizing, cleaning, and analyzing numerical data collected during research. Here are common methods used in the data processing phase of quantitative studies:

- Data entry
- Data cleaning
- Coding
- Data transformation
- Data aggregation
- Data validation
- Statistical analysis
- Software tools
- Data reduction
- Data interpretation
- Report writing
- Quality control

Data Entry:

- Transferring raw data from data collection instruments (such as surveys or questionnaires) to a digital format for analysis.
- Manual data entry or automated data entry using software tools.

Data Cleaning:

- Identifying and correcting errors, inconsistencies, and missing values in the dataset.
- Visual inspection of data.
- Statistical methods to detect outliers.
- Imputation of missing values.
- Validation checks to ensure data accuracy and consistency.

Coding:

- Assigning numerical codes to categorical variables for analysis.
- Developing a coding scheme and applying codes to transform qualitative data into a numerical format.

Data Transformation:

- Modifying or converting variables to meet the assumptions of statistical analyses or to create new variables.
- Logarithmic transformations.
- Standardization or normalization.
- Creating derived variables or aggregating data.

Data Aggregation:

- Combining individual data points into summary statistics or aggregated measures.
- Summation or averaging of values.
- Grouping data into categories or intervals.
- Creating frequency distributions.

Data Validation:

- Checking the dataset against predefined rules or criteria to ensure data quality.
- Range checks to identify values outside expected ranges.
- Cross-checking data for consistency.
- Comparison with external benchmarks or standards.

Statistical Analysis:

- Applying statistical techniques to analyze relationships, patterns, and trends in the data.
- Descriptive statistics (mean, median, mode, standard deviation).
- Inferential statistics (t-tests, chi-square tests, regression analysis).
- Multivariate analyses for more complex relationships.

Software Tools:

- Utilizing specialized software for efficient and accurate data processing and analysis.
- Spreadsheet software (Excel).
- Statistical software (SPSS, SAS).
- Database management systems (SQL).

Data Reduction:

- Simplifying the dataset by selecting relevant variables or summarizing data to focus on key aspects.
- Principal Component Analysis (PCA) for dimensionality reduction.
- Factor analysis to identify underlying factors.
- Selecting variables based on relevance to research objectives.

Data Interpretation:

- Drawing meaningful conclusions and interpretations from the analyzed data.
- Comparative analysis of results.
- Drawing inferences based on statistical significance.
- Linking findings to the research questions or hypotheses.

Report Writing:

- Communicating the results of the data analysis in a clear and concise manner.
- Presenting tables and graphs.
- Describing key findings and their implications.
- Providing limitations and recommendations for future research.

Quality Control:

- Implementing measures to ensure the accuracy, reliability, and validity of the data processing procedures.
- Peer review of data processing steps.
- Conducting data audits.
- Checking for consistency across different analytical techniques.

Methods of Data processing in Qualitative studies

Data processing in qualitative studies is iterative and often involves a flexible and reflexive approach to analysis. Researchers continually engage with the data, refine codes, and develop interpretations throughout the analysis process. The goal is to uncover rich, contextually embedded insights that contribute to a deeper understanding of the research phenomenon.

Data processing in qualitative studies involves the systematic organization, analysis, and interpretation of non-numerical data, often in the form of text, images, or audio. Here are common methods used in the data processing phase of qualitative studies:

- Data transcription
- Data coding
- Thematic analysis
- Content analysis
- Constant comparative analysis
- Narrative analysis
- Ground theory
- Matrix analysis
- Visual analysis
- Software tools

- Member checking
- Report writing

Data Transcription:

- Converting audio or video recordings into written text for analysis.
- Manual transcription or using automated transcription software.

Data Coding:

- Assigning labels or codes to segments of the data to identify patterns, themes, or concepts.
- Open coding- Initial coding without preconceived categories.
- Axial coding- Identifying relationships between codes.
- Selective coding- Refining and integrating codes into a coherent framework.

Thematic Analysis:

- Identifying, analyzing, and reporting patterns or themes within the qualitative data.
- In-depth reading of the data.
- Developing a coding framework based on emerging themes.
- Iteratively refining themes through data analysis.

Content Analysis:

- Systematically analyzing the content of textual, visual, or audio data to identify patterns or themes.
- Developing a coding scheme based on predefined categories.
- Coding data according to the established categories.
- Quantifying the frequency of coded categories.

Constant Comparative Analysis:

- Iteratively comparing new data with previously coded data to refine categories and concepts.
- Comparing new data with existing codes.
- Iteratively modifying codes and categories.
- Seeking patterns and variations across the dataset.

Narrative Analysis:

- Analyzing the structure and content of narratives to understand the stories and meanings conveyed by participants.
- Identifying key plot points and themes.
- Examining narrative elements such as characters, settings, and conflicts.

- Exploring the storytelling strategies employed.

Grounded Theory:

- Developing a theory based on systematically generated concepts from the data.
- Open coding to identify initial concepts.
- Constant comparison to refine concepts and relationships.
- Theoretical sampling to gather data that informs emerging theories.

Matrix Analysis:

- Organizing and comparing data using matrices or tables to facilitate systematic analysis.
- Developing a matrix with rows for cases and columns for themes or codes.
- Populating the matrix with coded data.
- Identifying patterns and relationships across cases and themes.

Visual Analysis:

- Analyzing visual data such as images, photographs, or videos to uncover patterns, meanings, or cultural insights.
- Describing visual elements.
- Analyzing the composition and context of visual data.
- Identifying themes or patterns within visual materials.

Software Tools:

- Using qualitative data analysis software to manage, code, and analyze large volumes of qualitative data.
- NVivo, MAXQDA, Dedoose, or other qualitative analysis tools.

Member Checking:

- Seeking feedback from participants to validate the accuracy and interpretation of findings.
- Sharing preliminary findings with participants.
- Incorporating participant feedback into the analysis.
- Enhancing the credibility and trustworthiness of the study.

Report Writing:

- Communicating the results of the qualitative analysis in a coherent and meaningful narrative.
- Organizing findings according to themes or concepts.
- Providing verbatim quotes to illustrate key points.
- Interpreting and discussing the implications of the findings.

Multiple Choice Questions

1. **What is the primary purpose of data coding in qualitative research?**

   a) To assign numerical values to qualitative data

   b) To summarize quantitative results

   c) To identify patterns, themes, or concepts in the data

   d) To conduct statistical analyses

   Answer: C) To identify patterns, themes, or concepts in the data

2. **In grounded theory coding, what is the initial stage where researchers assign labels to chunks of data without preconceived categories?**

   a) Axial coding

   b) Selective coding

   c) Open coding

   d) Constant comparative coding

   Answer: C) Open coding

3. **How does constant comparative analysis contribute to the coding process in qualitative research?**

   a) It finalizes coding categories.

   b) It compares qualitative and quantitative data.

   c) It iteratively refines codes and concepts.

   d) It automates the coding process.

   Answer: C) It iteratively refines codes and concepts

4. **What is the primary objective of data exploration in the research process?**

   a) To confirm pre-existing hypotheses

   b) To identify patterns, trends, and relationships in the data

   c) To collect new data for analysis

   d) To conduct statistical tests

   Answer: B) To identify patterns, trends, and relationships in the data

5. **Which statistical technique is commonly used in data exploration to visualize the distribution of a single variable?**

   a) Regression analysis

   b) Chi-square test

   c) Histograms

   d) T-tests

   Answer: C) Histograms

6. **In data exploration, what does a scatter plot help researchers visualize?**
   a) The distribution of a single variable
   b) The relationship between two variables
   c) The central tendency of the data
   d) The statistical significance of findings
   Answer: B) The relationship between two variables

7. **What is the primary goal of data editing in the research process?**
   a) To collect new data
   b) To analyze data trends
   c) To ensure data accuracy, consistency, and completeness
   d) To develop research hypotheses
   Answer: C) To ensure data accuracy, consistency, and completeness

8. **Which of the following is a common type of error addressed during data editing?**
   a) Hypothesis error b) Sampling error
   c) Transcription error d) Data analysis error
   Answer: C) Transcription error

9. **What does data cleaning involve in the context of research methodology?**
   a) Creating new variables
   b) Identifying and correcting errors in the dataset
   c) Conducting statistical analyses
   d) Developing data coding schemes
   Answer: B) Identifying and correcting errors in the dataset

10. **What is the primary purpose of data cleaning in quantitative research?**
    a) To collect new data
    b) To identify and correct errors in the dataset
    c) To develop research hypotheses
    d) To conduct statistical analyses
    Answer: B) To identify and correct errors in the dataset

11. **What statistical technique is commonly used for summarizing the central tendency of a dataset in quantitative studies?**
    a) Regression analysis b) Chi-square test
    c) Mean, Median, and Mode d) Analysis of Variance (ANOVA)
    Answer: C) Mean, Median, and Mode

12. In quantitative research, what does data transformation involve?

a) Converting categorical data into numerical data

b) Summarizing data using descriptive statistics

c) Identifying and correcting errors in the dataset

d) Modifying variables to meet the assumptions of statistical analyses

Answer: D) Modifying variables to meet the assumptions of statistical analyses

13. What is the primary purpose of data coding in qualitative research?

a) To assign numerical values to qualitative data

b) To summarize quantitative results

c) To identify patterns, themes, or concepts in the data

d) To conduct statistical analyses

Answer: C) To identify patterns, themes, or concepts in the data

14. Which qualitative analysis approach involves identifying, analyzing, and reporting patterns or themes within the data?

a) Content Analysis

b) Grounded Theory

c) Thematic Analysis

d) Matrix Analysis

Answer: C) Thematic Analysis

15. What does constant comparative analysis contribute to the coding process in qualitative research?

a) It finalizes coding categories.

b) It compares qualitative and quantitative data.

c) It iteratively refines codes and concepts.

d) It automates the coding process.

Answer: C) It iteratively refines codes and concepts

SHORT ANSWERS

1. What are the steps involved in data coding?
2. Write about data validation?
3. Define Data transcription?
4. What is Matrix analysis?
5. Define Grounded theory?
6. What are the Software tools involved in data processing in Quantitative studies?

Chapter 26

Quantitative Data Analysis

Analyzing number-based data (which includes categorical and numerical data) using various statistical techniques.

- Turn raw numbers into meaningful data through the application of rational and critical thinking.
- Calculation of frequencies of variables and differences between variables
- Associated with finding evidence to either support or reject hypotheses you have formulated at the earlier stages of your research process
- For example, questionnaire findings of a research titled "A study into the impacts of informal management-employee communication on the levels of employee motivation: a case study of Agro Bravo Enterprise" may indicate that the majority 52% of respondents assess communication skills of their immediate supervisors as inadequate.

What is Quantitative Analysis used for?

- Quantitative analysis is generally used for three purposes.
- Firstly, it's used to measure differences between groups. For example, the popularity of different clothing colors or brands.
- Secondly, it's used to assess relationships between variables. For example, the relationship between weather temperature and voter turnout.
- And third, it's used to test hypotheses in a scientifically rigorous way. For example, a hypothesis about the impact of a certain vaccine.

Parametric Data Analysis

Parametric statistics are **mathematical analyses of data that rely on strict assumptions about the underlying population parameters, usually for the purposes of comparing groups or variables.**

- Common parametric statistics, for example, **the Student's t-tests**
- Parametric statistics, the information about the distribution of the population is known and is based on a fixed set of parameters.
- When we use descriptive statistics it is useful to summarize our group of data using a combination of tabulated description (i.e., tables), graphical description (i.e., graphs and charts)

- Parametric statistical procedures rely on assumptions about the shape of the distribution (i.e., assume a normal distribution) in the underlying population and about the form or parameters (i.e., means and standard deviations) of the assumed distribution.

Common parametric tests include:

One sample t-test

Two sample t-test

One-way ANOVA

In order for the results of parametric tests to be valid, the following four assumptions should be met:

1. Normality – Data in each group should be normally distributed.
2. Equal Variance – Data in each group should have approximately equal variance.
3. Independence – Data in each group should be randomly and independently sampled from the population.
4. No Outliers – There should be no extreme outliers.

Descriptive Statistics

- Descriptive statistics is the term given to the analysis of data that helps describe, show or summarize data in a meaningful way such that, for example, patterns might emerge from the data.
- Descriptive statistics do not, however, allow us to make conclusions beyond the data.
- They are simply a way to describe our data.

For example,if we had the results of 100 pieces of students' coursework, we may be interested in the overall performance of those students. We would also be interested in the distribution or spread of the marks. Descriptive statistics allow us to do this.

There are two general types of statistic that are used to describe data (Descriptive)

1. Measures of central tendency :

This case, the frequency distribution is simply the distribution and pattern of marks scored by the 100 students from the lowest to the highest. We can describe this central position using a number of statistics, including the mode, median, and mean.

2. Measures of spread :

For example, the mean score of our 100 students may be 65 out of 100. However, not all students will have scored 65 marks. Rather, their scores will be spread out. Some will be lower and others higher. Measures of spread help us to summarize how spread out these scores

Inferential Statistics

- Inferential statistics is a branch of statistics that makes the use of various analytical tools to draw inferences about the population data from sample data. Apart from inferential statistics, descriptive statistics forms another branch of statistics. Inferential statistics help to draw conclusions about the population while descriptive statistics summarizes the features of the data set.
- There are two main types of inferential statistics - hypothesis testing and regression analysis. The samples chosen in inferential statistics need to be representative of the entire population. In this article, we will learn more about inferential statistics, its types, examples, and see the important formulas.

What Is Inferential Statistics?

Inferential statistics helps to develop a good understanding of the population data by analyzing the samples obtained from it. It helps in making generalizations about the population by using various analytical tests and tools. In order to pick out random samples that will represent the population accurately many sampling techniques are used. Some of the important methods are simple random sampling, stratified sampling, cluster sampling, and systematic sampling techniques.

Inferential Statistics Definition

Inferential statistics can be defined as a field of statistics that uses analytical tools for drawing conclusions about a population by examining random samples. The goal of inferential statistics is to make generalizations about a population. In inferential statistics, a statistic is taken from the sample data (e.g., the sample mean) that used to make inferences about the population parameter (e.g., the population mean).

Types of Inferential Statistics

Inferential statistics can be classified into hypothesis testing and regression analysis. Hypothesis testing also includes the use of confidence intervals to test the parameters of a population. Given below are the different types of inferential statistics.

Inferential Statistics

Inferential Statistics	
Hypothesis Testing	Regression Analysis
Z test	Linear Regression
F test	Nominal Regression
T test	Logistic Regression
ANOVA Test	Ordinal Regression
Wilcoxon Signed Rank Test	
Mann-Whitney U Test	

Hypothesis Testing

Hypothesis testing is a type of inferential statistics that is used to test assumptions and draw conclusions about the population from the available sample data. It involves setting up a null hypothesis and an alternative hypothesis followed by conducting a statistical test of significance. A conclusion is drawn based on the value of the test statistic, the critical value, and the confidence intervals. A hypothesis test can be left-tailed, right-tailed, and two-tailed. Given below are certain important hypothesis tests that are used in inferential statistics.

Z Test:

A z test is used on data that follows a normal distribution and has a sample size greater than or equal to 30. It is used to test if the means of the sample and population are equal when the population variance is known.

T Test:

A t test is used when the data follows a student t distribution and the sample size is lesser than 30. It is used to compare the sample and population mean when the population variance is unknown

F Test:

An f test is used to check if there is a difference between the variances of two samples or populations.

Anova Test

1. Also called as **Analysis of variance**, it is a parametric test of hypothesis testing.
2. It is an extension of the T-Test and Z-test.
3. It is used to test the significance of the differences in the mean values among more than two sample groups.

4. It uses F-test to statistically test the equality of means and the relative variance between them.

Confidence Interval

A confidence interval helps in estimating the parameters of a population. For example, a 95% confidence interval indicates that if a test is conducted 100 times with new samples under the same conditions then the estimate can be expected to lie within the given interval 95 times. Furthermore, a confidence interval is also useful in calculating the critical value in hypothesis testing.

Apart from these tests, other tests used in inferential statistics are the ANOVA test, Wilcoxon signed-rank test, Mann-Whitney U test, Kruskal-Wallis H test, etc.

Regression Analysis

Regression analysis is used to quantify how one variable will change with respect to another variable. There are many types of regressions available such as simple linear, multiple linear, nominal, logistic, and ordinal regression. The most commonly used regression in inferential statistics is linear regression. Linear regression checks the effect of a unit change of the independent variable in the dependent variable.

Inferential Statistics Examples

Inferential statistics is very useful and cost-effective as it can make inferences about the population without collecting the complete data. Some inferential statistics examples are given below:

- Suppose the mean marks of 100 students in a particular country are known. Using this sample information the mean marks of students in the country can be approximated using inferential statistics.
- Suppose a coach wants to find out how many average cartwheels sophomores at his college can do without stopping. A sample of a few students will be asked to perform cartwheels and the average will be calculated. Inferential statistics will use this data to make a conclusion regarding how many cartwheel sophomores can perform on average.

What are the Similarities Between Descriptive and Inferential Statistics?

- Both descriptive and inferential statistics rely on the same set of data. Descriptive statistics rely solely on this set of data, whilst inferential statistics also rely on this data in order to make generalisations about a larger population.
- Mostly, we use descriptive statistics for our analysis.
- What are the strengths of using descriptive statistics to examine a distribution of scores?

- Other than the clarity with which descriptive statistics can clarify large volumes of data, there are no uncertainties about the values you get (other than only measurement error, etc.)

Non Parametric Statistics

Nonparametric statistics refers to a statistical method in which the data are not assumed to come from prescribed models that are determined by a small number of parameters.

Also known as distribution-free tests because they are based on fewer assumptions (e.g., they do not assume that the outcome is approximately normally distributed).

Examples of such models include the normal distribution model and the linear regression model.

Types Non Parametric Test

- Chi-Square Test
- Mann-Whitney U-Test
- Kruskal-Wallis H-test

Chi-Square Test

- It is a non-parametric test of hypothesis testing.
- As a non-parametric test, chi-square can be used:
  - test of goodness of fit.
  - as a test of independence of two variables.
- It helps in assessing the goodness of fit between a set of observed and those expected theoretically.
- It makes a comparison between the expected frequencies and the observed frequencies.

Mann-Whitney U-Test

- It is a non-parametric test of hypothesis testing.
- This test is used to investigate whether two independent samples were selected from a population having the same distribution.
- It is a true non-parametric counterpart of the T-test and gives the most accurate estimates of significance especially when sample sizes are small and the population is not normally distributed.
- It is based on the comparison of every observation in the first sample with every observation in the other sample.

Kruskal-Wallis H-test

- It is a non-parametric test of hypothesis testing.

- ❖ This test is used for comparing two or more independent samples of equal or different sample sizes.
- ❖ It extends the Mann-Whitney-U-Test which is used to comparing only two groups.

Reasons to Use Non-Parametric Tests

It is important to access when to apply parametric and non-parametric tests in order to arrive at the correct statistical inference. The reasons to use a non-parametric test are given below:

- ❖ When the distribution is skewed, a non-parametric test is used. For skewed distributions, the mean is not the best measure of central tendency, hence, parametric tests cannot be used.
- ❖ If the size of the data is too small then validating the distribution of the data becomes difficult. Thus, in such cases, a non-parametric test is used to analyze the data.
- ❖ If the data is nominal or ordinal, a non-parametric test is used. This is because a parametric test can only be used for continuous data.

Important Notes on Non-Parametric Test

- ❖ A non-parametric test is a statistical test that is performed on data belonging to a distribution whose parameters are unknown.
- ❖ It is used on skewed distributions and the measure of central tendency used is the median.
- ❖ Kruskal Wallis test, sign test, Wilcoxon signed test and the Mann Whitney u test are some important non-parametric tests used in hypothesis testing.

Key Difference Between Parametric and Non Parametric Test

BASIS FOR COMPARISON	*PARAMETRIC TEST*	*NONPARAMETRIC TEST*
Meaning	A statistical test, in which specific assumptions are made about the population parameter is known as parametric test.	A statistical test used in the case of non-metric independent variables, is called non-parametric test.
Basis of test statistic	Distribution	Arbitrary
Measurement level	Interval or ratio	Nominal or ordinal
Measure of central tendency	Mean	Median
Information about population	Completely known	Unavailable
Applicability	Variables	Variables and Attributes

Hypothesis Testing

- Hypothesis testing is an act in statistics whereby an analyst tests an assumption regarding a population parameter.
- Hypothesis testing in statistics is a way for you to test the results of a survey or experiment to see if you have meaningful results. You're basically testing whether your results are valid by figuring out the odds that your results have happened by chance
- For example, a person wants to test that a penny has exactly a 50% chance of landing on heads, the null hypothesis would be that 50% is correct, and the alternative hypothesis would be that 50% is not correct.

$$z = \frac{\hat{p} - p}{\sqrt{p\,q/n}}$$

Five Steps of Hypothesis Testing:

- Specify the Null Hypothesis.
- Specify the Alternative Hypothesis.
- Set the Significance Level (a)
- Calculate the Test Statistic and Corresponding P-Value.
- Drawing a Conclusion.

Type I and II Errors

- A type I error (false-positive) occurs if an investigator rejects a null hypothesis that is actually true in the population; a type II error (false-negative) occurs if the investigator fails to reject a null hypothesis that is actually false in the population.
- Although type I and type II errors can never be avoided entirely, the investigator can reduce their likelihood by increasing the sample size (the larger the sample, the lesser is the likelihood that it will differ substantially from the population).
- Example: Type I vs Type II error
- You decide to get tested for COVID-19 based on mild symptoms. There are two errors that could potentially occur:
- **Type I error (false positive): the test result says you have coronavirus, but you actually don't.**
- Type II error (false negative): the test result says you don't have coronavirus, but you actually do.

Type I and Type II Error

Null hypothesis is ...	True	False
Rejected	Type I error False positive Probability = α	Correct decision True positive Probability = 1 - β
Not rejected	Correct decision True negative Probability = 1 - α	Type II error False negative Probability = β

Scribbr

References

https://www.healthknowledge.org.uk/public-health-textbook/research-methods/1b-statistical-methods/parametric-nonparametric-tests#:~:text=Parametric%20tests%20are%20those%20that,used%20for%20non%2DNormal%20variables

https://ebooks.inflibnet.ac.in/hsp16/chapter/parametric-and-non-parametric-test-meaning-importance-application-merits-and-demerits/

https://www.scribbr.com/statistics/type-i-and-type-ii-errors/

https://en.wikipedia.org/wiki/Type_I_and_type_II_errors

Question

1. What is Quantitative data analysis?
2. What is the purpose of Quantitative analysis?
3. Explain – Parametric Data Analysis.
4. What are the similarities between Descriptive and Inferential statistics?
5. Explain – Non Parametric Data Analysis.
6. Difference between Parametric Data Analysis - Non Parametric Data Analysis.
7. How to testing the Hypothesis?
8. What all are the steps in testing the Hypothesis?
9. Tell about Type I and II Errors.
10. Give any example of Type I and II Errors.

Fill Ups

1. _______ is the data analysis method that analyzing number-based data.
2. Population parameters, usually for the purpose of comparing _______

3. A Example for Parametric Statistics _______
4. _______ is the term given to the analysis of data that helps describe, show summarize data in meaningful way.
5. _______ are the techniques that allow us to use these samples to make generalizations about the populations from which the samples were drawn.
6. Example for Non Parametric Statistics_______
7. _______ is an act in statistics whereby an analyst tests on assumption regarding population parameter.
8. Which Error is refered as False-positive error______
9. ______ is refered as False-negative error.
10. ______ statistics do not, however, allow us to make conclusion beyond the data.

Chapter 27

Hypothesis Testing and Multivariate Analysis

Hypothesis

A hypothesis is a statement that introduces a research question and proposes an expected result. It is an integral part of the scientific method that forms the basis of scientific experiments

Hypothesis Testing

- ❖ Hypothesis testing is an act in statistics whereby an analyst tests an assumption regarding a population parameter. The methodology employed by the analyst depends on the nature of the data used and the reason for the analysis. Hypothesis testing is used to assess the possibility of a hypothesis by using sample data.
- ❖ The formal procedure statisticians follow to determine whether a certain hypothesis is valid or not is referred to as hypothesis testing.
- ❖ Hypothesis testing is a 4-step process
  - ▷ Write the hypothesis
  - ▷ Create an analysis plan
  - ▷ Analyse the data
  - ▷ Interpret the data

Method Of Hypothesis Testing

- ❖ To begin, we identify a hypothesis or claim that we feel should be tested. For example, we might want to test the claim that the mean number of hours that children in the United States watch TV is 3 hours.
- ❖ We select a criterion upon which we decide that the claim being tested is true or not. For example, the claim is that children watch 3 hours of TV per week. Most samples we select should have a mean close to or equal to 3 hours if the claim we are testing is true. So at what point do we decide that the discrepancy between the sample mean and 3 is so big that the claim we are testing is likely, not true? We answer this question in this step of hypothesis testing.

- Select a random sample from the population and measure the sample mean. For example, we could select 20 children and measure the mean time (in hours) that they watch TV per week.
- Compare what we observe in the sample to what we expect to observe if the claim we are testing is true. We expect the sample mean to be around 3 hours. If the discrepancy between the sample mean and population mean is small, then we will likely decide that the claim we are testing is indeed true. If the discrepancy is too large, then we will likely decide to reject the claim as being not true.

Concepts Of Hypothesis Testing

It consists of several components; two statements,

- The null hypothesis and the alternative hypothesis,
- The test statistic and the crucial value, which in tern give us the p- value and the rejection value respectively.

Effect Size

- Effect size is a statistical concept that measures the strength of the relationship between two variables on a numeric scale.
- In hypothesis testing, effect size is the objective and standardized measure of the size of a particular effect.
- The effect size tells us how much impact our effect has on our test population. The larger the effect size, the more important the effect.
- Examples of effect sizes include the correlation between two variables, the regression coefficient in a regression, the mean difference, or the risk of a particular event happening.

α (Alpha):

- Alpha, the significance level, is the probability that you will make the mistake of rejecting the null hypothesis when in fact it is true.
- α (Alpha) is the probability of Type I error in any hypothesis test– incorrectly rejecting the null hypothesis.
- For example, if you want to be 95 percent confident that your analysis is correct, the alpha level would be 1–. 95 = 5 percent, assuming you had a one tailed test.

β (Beta)

- Its is the probability of Type II error in any hypothesis test–incorrectly failing to reject the null hypothesis. (1 – β is power)

Power P Value:

- The p-value measures the probability of getting a more extreme value than the one you got from the experiment. If the p-value is greater than alpha, you accept the null hypothesis

Power = 1 – β

A p-value is a probability of obtaining a sample outcome, given that the value stated in the null hypothesis is true. The p-value for obtaining a sample outcome is compared to the level of significance.

- ❖ The p-value (or p-value or probability value) is the probability of getting a value of the test statistic that is at least as extreme as the one representing the sample data, assuming that the null hypothesis is true.
- ❖ The null hypothesis is rejected if the P-value is very small, such as 0.05 or less.
- ❖ Significance, or statistical significance, describes a decision made concerning a value stated in the null hypothesis. When the null hypothesis is rejected, we reach significance. When the null hypothesis is retained, we fail to reach significance.
- ❖ When the p-value is less than 5% ($p < 0.05$), we reject the null hypothesis. We will refer to $p < 0.05$ as the criterion for deciding to reject the null hypothesis, although note that when $p = 0.05$, the decision is also to reject the null hypothesis. When the p-value is greater than 5% ($p > 0.05$), we retain the null hypothesis. The decision to reject or retain the null hypothesis is called **significance.**
- ❖ When the p-value is less than 0.05, we reach significance; the decision is to reject the null hypothesis. When the p-value is greater than 0.05, we fail to reach significance; the decision is to retain the null hypothesis.

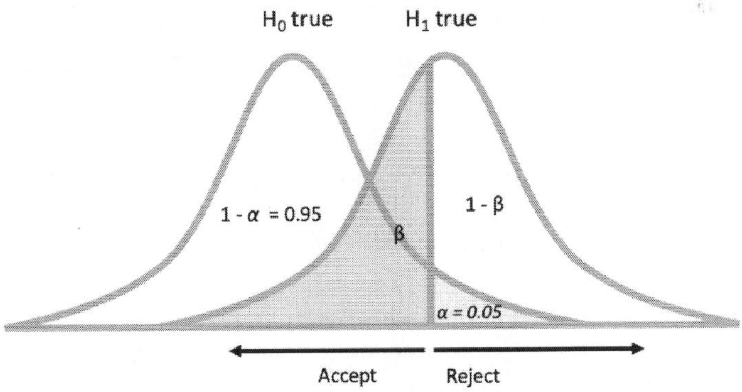

Factor Analysis

- ❖ Factor analysis is a way to condense the data in many variables into just a few variables. It is also called as **"dimension reduction".**
- ❖ It makes the grouping of variables with high correlation. It includes techniques such as principal component analysis and common factor analysis.
- ❖ This type of technique is used as a pre-processing step to transform the data before using other models.

- When the data has too many variables, the performance of multivariate techniques is not at the optimum level, as patterns are more difficult to find.
- By using factor analysis, the patterns become less diluted and easier to analyse.

Multivariate Data Analysis

- Multivariate Analysis is defined as a process of involving multiple dependent variables resulting in one outcome. This explains that the majority of the problems in the real world are Multivariate.
- For example, we cannot predict the weather of any year based on the season. There are multiple factors like pollution, humidity, precipitation, etc. Here, we will introduce you to multivariate analysis, its history, and its application in different fields.

Cluster Analysis

- Cluster analysis is a class of techniques that are used to classify objects or cases into relative groups called clusters. In cluster analysis, there is no prior information about the group or cluster membership for any of the objects.
- While doing cluster analysis, we first partition the set of data into groups based on data similarity and then assign the labels to the groups.
- The main advantage of clustering over classification is that it is adaptable to changes and helps single out useful features that distinguish different groups.
- Cluster Analysis used in outlier detection applications such as detection of credit card fraud. As a data mining function, cluster analysis serves as a tool to gain insight into the distribution of data to observe the characteristics of each cluster.

Logistic Regression

- A regression analysis used when the dependent variable is categorical. It predicts the probability of a categorical outcome based on predictor variables.
- Logistic regression is used to obtain odds ratio in the presence of more than one explanatory variable. The procedure is quite similar to multiple linear regression, with the exception that the response variable is binomial. The result is the impact of each variable on the odds ratio of the observed event of interest.
- For example, a logistic regression could be used to predict whether a political candidate will win or lose an election or whether a high school student will be admitted or not to a particular college. These binary outcomes allow straightforward decisions between two alternatives.

Structural Equation Modelling

- A statistical method that examines the relationships between multiple variables. It's used to test and estimate complex relationships and causal pathways among variables.

- Structural equation modelling is **a multivariate statistical analysis technique that is used to analyse structural relationships.**
- It is an extremely broad and flexible framework for data analysis, perhaps better thought of as a family of related methods rather than as a single technique.
- SEM in a single analysis can assess the assumed causation among a set of dependent and independent constructs i.e. validation of the structural model and the loadings of observed items (measurements) on their expected latent variables (constructs) i.e. validation of the measurement model.
- The combined analysis of the measurement and the structural model enables the measurement errors of the observed variables to be analysed as an integral part of the model, and factor analysis combined in one operation with the hypotheses testing.

Guidelines for Choosing Appropriate Statistical Analysis

Three criteria are decisive for the selection of the statistical analysis, which are as follows:

- The number of variables,
- Types of data/level of measurement (continuous, binary, categorical) and
- The type of study design (paired or unpaired).

The Number of Variables that the Test is to be Conducted on

- Statistical tests and procedures can be divided according to the number of variables that they are designed to analyse.
- Therefore, when choosing a test it is important that you consider how many variables one wishes to analyse.
- One set of tests is used on single variables (often referred to as descriptive statistics), a second set is used to analyse the relationship between two variables and a third set used to model multivariable relationships (i.e., relationships between three or more variables).
- Types of data: Continuous, categorical, or binary
- Categorical variables are race, sex, age group, and educational level.
- Continuous data is data that can take any value. Height, weight, temperature and length are all examples of continuous data.
- A binary variable (such as yes/no question) is a categorical variable having two categories (yes or no) and there is no intrinsic ordering to the categories.

Paired and Unpaired Study Designs

- A paired t-test (also known as a dependent or correlated t-test) is a statistical test that compares the averages/means and standard deviations of two related groups to determine if there is a significant difference between the two groups.

- There are two possible hypotheses in a paired t-test (Null hypothesis (H0) & alternative hypothesis (H1)
- An unpaired t-test (also known as an independent t-test) is a statistical procedure that compares the averages/means of two independent or unrelated groups to determine if there is a significant difference between the two.

Multiple Choice Questions

1. **Which of the following is true about a null hypothesis (H0)?**

   a) It states the alternative viewpoint

   b) It's the hypothesis researchers want to support

   c) It assumes no effect or no difference

   d) It's always accepted in hypothesis testing

2. **What is the p-value in hypothesis testing?**

   a) The probability of observing the sample data given that the null hypothesis is true

   b) The probability of rejecting the null hypothesis when it's true

   c) The probability of accepting the null hypothesis when it's true

   d) The probability of obtaining the same sample data in future experiments

3. **A Type I error occurs when:**

   a) The null hypothesis is true, but rejected

   b) The null hypothesis is false, but accepted

   c) The alternative hypothesis is true, but rejected

   d) The alternative hypothesis is false, but accepted

4. **The significance level (alpha) in hypothesis testing represents:**

   a) The probability of making a Type II error

   b) The probability of rejecting the null hypothesis when it's true

   c) The acceptable level of Type I error

   d) The probability of accepting the null hypothesis when it's false

5. **What is the critical region in hypothesis testing?**

   a) The region where the null hypothesis is rejected

   b) The area where the alternative hypothesis holds true

   c) The area where the sample mean is highest

   d) The region where Type II errors occur

6. **The null hypothesis is rejected if the test statistic is:**

   a) Greater than the critical value

   b) Less than the critical value

   c) Equal to the critical value

   d) Not related to the critical value

7. **Which value determines the sample size needed for a hypothesis test?**

   a) Power

   b) Significance level

   c) Effect size

   d) Type I error rate

8. **Which statement is correct regarding a confidence interval?**

   a) It indicates the range of population values that includes the sample statistic

   b) It's equivalent to a p-value

   c) It's used to determine the effect size

   d) It's independent of the sample size

9. **The critical region in hypothesis testing is defined by:**

   a) The p-value

   b) The test statistic

   c) The significance level

   d) The samplemean

10. **What does a large p-value indicate in hypothesis testing?**

    a) Strong evidence against the null hypothesis

    b) Weak evidence against the null hypothesis

    c) Acceptance of the null hypothesis

    d) An error in the analysis

Definition

1. Define Null hypothesis.
2. Define Alternative hypothesis
3. Define significance level
4. Define Hypothesis testing
5. Define P- value
6. Define SEM
7. Define Cluster analysis
8. Define Logistic regression

Chapter 28

Statistical Software for Social Research

Introduction

Statistical packages are essential tools in the field of social research, providing researchers with powerful capabilities to analyze and interpret data. These software applications are designed to handle the complexities of statistical analysis, allowing researchers to explore patterns, relationships, and trends within datasets. They play a crucial role in transforming raw data into meaningful insights, aiding researchers in testing hypotheses, making inferences, and presenting findings.

Among the notable statistical packages used in social research are SPSS (Statistical Package for the Social Sciences), R, Stata, SAS (Statistical Analysis System), MATLAB, Python with libraries like Pandas and NumPy, Excel, IBM SPSS Modeler, and JASP (Jeffreys's Amazing Statistics Program). Each package offers a unique set of features and capabilities, catering to the diverse needs of researchers across various disciplines within the social sciences

Researchers utilize these packages for tasks such as descriptive statistics, inferential statistics, data visualization, regression analysis, and predictive modeling. The choice of a specific statistical package often depends on factors like the complexity of the research questions, the nature of the data, and the researcher's familiarity with the software. As technology advances, the integration of these packages with machine learning techniques further enhances their utility in addressing complex research challenges.

Statistical packages serve as indispensable tools for social researchers, enabling them to navigate the intricate landscape of data analysis and derive meaningful conclusions from empirical evidence. Their versatility, coupled with a diverse array of statistical functions, empowers researchers to conduct rigorous and sophisticated analyses, contributing to advancements in our understanding of social phenomena.

Some of the statistical packages are listed below,

- SPSS (Statistical Package for the Social Sciences)
- SAS (Statistical Analysis System)
- R

- Python with Pandas and Statsmodels
- Stata
- IBM SPSS Statistics
- JMP
- MATLAB
- HLM (Hierarchical Linear Modeling)
- NVivo
- Atlas.ti
- Mplus
- EQS (Equation Modeling Software)
- WinBUGS/OpenBUGS/JAGS
- UCINet
- GRETL (Gnu Regression, Econometrics, and Time-series Library)
- PSPP

SPSS (Statistical Package for the Social Sciences):

- *Introduction:* SPSS is a widely used statistical software known for its user-friendly interface and extensive capabilities in social research.
- *Statistical Functions:*
  - Descriptive Statistics: Mean, median, mode, standard deviation.
  - Inferential Statistics: T-tests, ANOVA, regression analysis.
  - Factor Analysis and Cluster Analysis.
  - Data Visualization: Various chart options.

SAS (Statistical Analysis System):

- *Introduction:* SAS is a powerful statistical software suite used for advanced analytics and data management.
- *Statistical Functions:*
  - Data Management: Importing, cleaning, and organizing datasets.
  - Advanced Analytics: Multivariate analysis, survival analysis.
  - Inferential Statistics: Hypothesis testing, regression analysis.

R:

- *Introduction:* R is a free and open-source programming language and software environment for statistical computing.
- *Statistical Functions:*
  - Descriptive Statistics: Summary statistics, measures of central tendency.
  - Inferential Statistics: T-tests, chi-square tests, regression analysis.

- ▷ Data Visualization: Various plots and charts.
- ▷ Machine Learning: Classification, regression, clustering.

Python with Pandas and Statsmodels:

- ❖ *Introduction:* Python is a versatile programming language with Pandas for data manipulation and Statsmodels for statistical analysis.
- ❖ *Statistical Functions:*
  - ▷ Data Management: Handling and transforming datasets using Pandas.
  - ▷ Descriptive Statistics: Calculating summary statistics.
  - ▷ Regression Analysis: Implementing linear and logistic regression using Statsmodels.
  - ▷ Machine Learning: Applying scikit-learn for classification, regression, clustering.

Stata:

- ❖ *Introduction:* Stata is a statistical software package known for its user-friendly interface and versatility.
- ❖ *Statistical Functions:*
  - ▷ Data Management: Importing, cleaning, and managing datasets.
  - ▷ Inferential Statistics: Hypothesis testing, ANOVA, regression analysis.
  - ▷ Time Series Analysis: Modeling and analyzing time-dependent data.
  - ▷ Multivariate Analysis: MANOVA, discriminant analysis.

IBM SPSS Statistics:

- ❖ *Introduction:* IBM SPSS Statistics is an advanced version of SPSS, offering additional features.
- ❖ Statistical Functions:
  - ▷ Descriptive Statistics: Summary measures, frequency distributions.
  - ▷ Inferential Statistics: T-tests, ANOVA, regression analysis.
  - ▷ Advanced Analytics: Predictive modeling, decision trees.

JMP:

- ❖ *Introduction:* JMP is a statistical software for dynamic data visualization and exploratory data analysis.
- ❖ *Statistical Functions:*
  - ▷ Data Visualization: Dynamic and interactive visualization.
  - ▷ Inferential Statistics: Hypothesis testing, ANOVA, regression.

MATLAB:

- ❖ *Introduction:* MATLAB is a programming language for numerical computing and statistical analysis.

- *Statistical Functions:*
  - ▷ Descriptive Statistics: Mean, median, standard deviation.
  - ▷ Inferential Statistics: Hypothesis testing, regression.
  - ▷ Time Series Analysis: Modeling and simulating time-series data.

HLM (Hierarchical Linear Modeling):

- *Introduction:* HLM is specialized for hierarchical and multilevel modeling, commonly used in educational and social research.
- *Statistical Functions:*
  - ▷ Hierarchical Modeling: Analyzing data with nested structures.
  - ▷ Multilevel Analysis: Modeling relationships at different levels.
  - ▷ Random Effects Modeling: Incorporating random effects in analyses.

NVivo:

- *Introduction:*NVivo is qualitative data analysis software used for coding, organizing, and analyzing textual, visual, and audio data.
- *Statistical Functions:*
  - ▷ Qualitative Coding: Organizing and categorizing textual and multimedia data.
  - ▷ Theme Extraction: Identifying recurring themes in qualitative data.

Atlas.ti:

- *Introduction:*Atlas.ti is qualitative data analysis software, facilitating the coding and analysis of textual, visual, and audio data.
- *Statistical Functions:*
  - ▷ Qualitative Coding: Coding and categorizing qualitative data.
  - ▷ Content Analysis: Analyzing textual, visual, and audio data for themes.

Mplus:

- *Introduction:*Mplus specializes in structural equation modeling (SEM) and latent variable analysis.
- *Statistical Functions:*
  - ▷ SEM: Analyzing complex relationships and latent variables.
  - ▷ Latent Variable Analysis: Estimating relationships between unobserved variables.

EQS (Equation Modeling Software):

- *Introduction:* EQS is designed for structural equation modeling (SEM) and covariance structure analysis.

- *Statistical Functions:*
  - SEM: Modeling complex relationships and pathways.
  - Covariance Structure Analysis: Examining relationships in multivariate data.

WinBUGS/OpenBUGS/JAGS:

- *Introduction:* These are Bayesian statistical software tools for modeling complex systems using Bayesian inference.
- *Statistical Functions:*
  - Bayesian Data Analysis: Implementing Bayesian statistical models.
  - Markov Chain Monte Carlo (MCMC): Sampling methods for Bayesian inference.

UCINet:

- *Introduction:*UCINet is software for social network analysis, used to analyze and visualize relationships within networks.
- *Statistical Functions:*
  - Network Analysis: Examining relationships, centrality measures.
  - Visualization: Representing network structures visually.

GRETL (Gnu Regression, Econometrics, and Time-series Library):

- *Introduction:* GRETL is an econometrics package for statistical modeling and analysis.
- *Statistical Functions:*
  - Time-Series Analysis: Modeling and forecasting economic and time-series data.
  - Econometric Modeling: Analyzing relationships in economic data.

PSPP:

- *Introduction:* PSPP is an open-source alternative to SPSS, providing basic statistical functions.
- *Statistical Functions:*
  - Descriptive Statistics: Mean, median, mode.
  - Inferential Statistics: T-tests, ANOVA, regression analysis.

Importance of Statistical Packages in Data Analyses

Statistical packages play a crucial role in data analysis for social research, offering a range of tools and functionalities that enhance the research process. Here are key aspects highlighting the importance of statistical packages in social research:

1. Efficient Data Management:

- ❖ Statistical packages facilitate the import, cleaning, and organization of large datasets, ensuring efficient data management.
- ❖ Researchers can handle missing data, transform variables, and prepare datasets for analysis using these tools.

2. Descriptive Statistics:

- ❖ Statistical packages provide a comprehensive set of descriptive statistics, including measures of central tendency, dispersion, and frequency distributions.
- ❖ Descriptive statistics offer insights into the characteristics of the data, aiding researchers in understanding the distribution and variability of variables.

3. Inferential Statistics:

- ❖ Researchers can apply a variety of inferential statistical techniques, such as hypothesis testing, ANOVA, and regression analysis, to make predictions and draw conclusions about populations based on sample data.
- ❖ These methods are fundamental for testing research hypotheses and establishing the significance of relationships within social phenomena.

4. Advanced Analytical Techniques:

- ❖ Statistical packages support advanced analytical methods like factor analysis, cluster analysis, and structural equation modeling.
- ❖ These techniques enable researchers to uncover complex relationships, identify latent variables, and model intricate structures within social data.

5. Data Visualization:

- ❖ Visualization tools within statistical packages allow researchers to create charts, graphs, and plots that enhance the communication of research findings.
- ❖ Visual representation aids in interpreting trends, patterns, and relationships within the data, making it accessible to a wider audience.

6. Interdisciplinary Applications:

- ❖ Statistical packages find applications across various social science disciplines, including sociology, psychology, economics, political science, and more.
- ❖ Researchers can tailor their analyses to suit the specific methodologies and research questions relevant to their fields.

7. Qualitative Data Analysis:

- ❖ Some statistical packages integrate with qualitative data analysis software, allowing researchers to combine quantitative and qualitative insights.

- This integration enhances the depth and richness of social research by incorporating both numerical and textual data.

8. Time-Series Analysis:

- Statistical packages support time-series analysis, allowing researchers to model and analyze temporal patterns and trends in social data.
- Time-series techniques are essential for understanding the dynamics of social phenomena over time.

9. Statistical Learning and Machine Learning:

- Some packages incorporate statistical learning and machine learning algorithms, enabling researchers to develop predictive models and uncover complex patterns in social datasets.
- Machine learning techniques enhance the predictive capabilities of social research models.

10. Standardization and Reproducibility:

- Statistical packages provide a standardized platform for conducting analyses, ensuring that researchers adhere to established statistical methods.
- The reproducibility of analyses is facilitated, allowing other researchers to replicate and verify results.

In summary, statistical packages are indispensable tools in social research, offering a systematic and efficient way to analyze data, test hypotheses, and derive meaningful insights. Their versatility, user-friendly interfaces, and support for diverse statistical methods contribute significantly to advancing knowledge in the social sciences.

Multiple Choice Questions

1. **Question: Which statistical package is specifically designed for social science research?**

   a) R

   b) Stata

   c) MATLAB

   d) Excel

   Answer: B) Stata

2. **Question: In R, what does the term "data frame" refer to?**

   a) A type of statistical test

   b) A graphical representation

   c) A data structure for storing datasets

   d) A regression model

   Answer: C) A data structure for storing datasets

3. **Question: Which programming language is commonly associated with the statistical package Python for data analysis?**

   a) Java

   b) C++

   c) Ruby

   d) Python

   Answer: D) Python

4. **Question: What is the primary purpose of the "Descriptive Statistics" module in SPSS?**

   a) Conducting hypothesis tests

   b) Cleaning and transforming data

   c) Summarizing and describing data

   d) Building predictive models

   Answer: C) Summarizing and describing data

5. **Question: Which statistical package is known for its emphasis on Bayesian statistics?**

   a) SAS

   b) JASP

   c) Excel

   d) IBM SPSS Modeler

   Answer: B) JASP

6. **Question: In MATLAB, what does the "plot" function do?**

   a) Perform hypothesis tests

   b) Create histograms

   c) Generate descriptive statistics

   d) Create visual plots and graphs

   Answer: D) Create visual plots and graphs

7. **Question: Which statistical package is commonly used for econometric analysis and time-series analysis?**

   a) R

   b) Stata

   c) SAS

   d) Excel

   Answer: B) Stata

8. **Question: What is the primary role of the "ANOVA" function in statistical packages?**

a) Descriptive statistics

b) Analysis of variance

c) Regression analysis

d) Data visualization

Answer: B) Analysis of variance

9. **Question: Which statistical package has a specific module called "Modeler" for predictive analytics?**

a) R

b) SAS

c) IBM SPSS Modeler

d) Python

Answer: C) IBM SPSS Modeler

10. **Question: In SPSS, what does the term "Factor Analysis" refer to?**

a) A type of regression analysis

b) A method for clustering data

c) A technique for reducing data dimensionality

d) A form of Bayesian analysis

Answer: C) A technique for reducing data dimensionality

11. **Question: Which statistical package is often used for machine learning applications through its scikit-learn library?**

a) Stata

b) R

c) Python

d) JASP

Answer: C) Python

12. **Question: What does the acronym "SPSS" stand for?**

a) Statistical Program for Social Sciences

b) Scientific Programming and Statistical System

c) Social and Predictive Statistical Software

d) Systematic Programming for Statistical Solutions

Answer: A) Statistical Program for Social Sciences

13. **Question: Which statistical package is commonly associated with the term "CRAN" (Comprehensive R Archive Network)?**

    a) SAS

    b) SPSS

    c) R

    d) JASP

    Answer: C) R

14. **Question: In JASP, what does the term "Bayesian Factor" represent?**

    a) A type of correlation coefficient

    b) A measure of evidence for hypotheses

    c) A method for data imputation

    d) A type of regression analysis

    Answer: B) A measure of evidence for hypotheses

15. **Question: Which statistical package is often used for survey data analysis due to its specialized features?**

    a) MATLAB

    b) SPSS

    c) Excel

    d) JASP

    Answer: B) SPSS

Reference

"Research Methodology: Methods and Techniques" by C.R. Kothari

Tests, Measurements and Research Methods in Behavioural Sciences by Arun Kumar Singh

https://en.m.wikipedia.org

Chapter 29

Interpretation and Presentation of Data

Data Interpretation

Data interpretation is the process of reviewing data through some predefined processes which will help assign some meaning to the data and arrive at a relevant conclusion. It involves taking the result of data analysis, making inferences on the relations studied, and using them to conclusion. Data analysis is the process of ordering, categorizing, manipulating, and summarizing data to obtain answers to research questions. It is usually the first step taken towards data interpretation.

It is evident that the interpretation of data is very important, and as such needs to be done properly. Therefore, researchers have identified some data interpretation methods to aid this process.

Methods of Data Interpretation

Data interpretation methods are how analysts help people make sense of numerical data that has been collected, analyzed and presented. Data, when collected in raw form, may be difficult for the layman to understand, which is why analysts need to break down the information gathered so that others can make sense of it.

For example, when founders are pitching to potential investors, they must interpret data (e.g. market size, growth rate, etc.) for better understanding. There are 2 main methods in which this can be done, namely; quantitative methods and qualitative methods.

Qualitative Data Interpretation Method

The qualitative data interpretation method is used to analyze qualitative data, which is also known as categorical data. With this type of analysis, data is not described through numerical values or patterns, but through the use of descriptive context (i.e., text). Typically, narrative data is gathered by employing a wide variety of person-to-person techniques. These techniques include:

- **Observations:** detailing behavioral patterns that occur within an observation group. These patterns could be the amount of time spent in an activity, the type of activity, and the method of communication employed.

- **Focus groups:** Group people and ask them relevant questions to generate a collaborative discussion about a research topic.
- **Secondary Research:** much like how patterns of behavior can be observed, various types of documentation resources can be coded and divided based on the type of material they contain.
- **Interviews:** one of the best collection methods for narrative data. Inquiry responses can be grouped by theme, topic, or category. The interview approach allows for highly-focused data segmentation.

A key difference between qualitative and quantitative analysis is clearly noticeable in the interpretation stage. The first one is widely open to interpretation and must be "coded" so as to facilitate the grouping and labeling of data into identifiable themes. As person-to-person data collection techniques can often result in disputes pertaining to proper analysis, qualitative data analysis is often summarized through three basic principles: notice things, collect things, and think about things.

After qualitative data has been collected through transcripts, questionnaires, audio and video recordings, or the researcher's notes, it is time to interpret it. For that purpose, there are some common methods used by researchers and analysts.

- **Content analysis**: As its name suggests, this is a research method used to identify frequencies and recurring words, subjects and concepts in image, video, or audio content. It transforms qualitative information into quantitative data to help in the discovery of trends and conclusions that will later support important research or business decisions. This method is often used by marketers to understand brand sentiment from the mouths of customers themselves. Through that, they can extract valuable information to improve their products and services. It is recommended to use content analytics tools for this method as manually performing it is very time-consuming and can lead to human error or subjectivity issues. Having a clear goal in mind before diving into it is another great practice for avoiding getting lost in the fog.
- **Thematic analysis:** This method focuses on analyzing qualitative data such as interview transcripts, survey questions, and others, to identify common patterns and separate the data into different groups according to found similarities or themes. For example, imagine you want to analyze what customers think about your restaurant. For this purpose, you do a thematic analysis on 1000 reviews and find common themes such as "fresh food", "cold food", "small portions", "friendly staff", etc. With those recurring themes in hand, you can extract conclusions about what could be improved or enhanced based on your customer's experiences. Since this technique is more exploratory, be open to changing your research questions or goals as you go.
- **Narrative analysis:** A bit more specific and complicated than the two previous methods, narrative analysis is used to analyze stories and discover the meaning behind them. These stories can be extracted from testimonials,

case studies, and interviews as these formats give people more space to tell their experiences. Given that collecting this kind of data is harder and more time-consuming, sample sizes for narrative analysis are usually smaller, which makes it harder to reproduce its findings. However, it still proves to be a valuable technique in cases such as understanding customers' preferences and mindsets.

- **Discourse analysis**: This method is used to draw the meaning of any type of visual, written, or symbolic language in relation to a social, political, cultural, or historical context. It is used to understand how context can affect the way language is carried out and understood. For example, if you are doing research on power dynamics, using discourse analysis to analyze a conversation between a janitor and a CEO and draw conclusions about their responses based on the context and your research questions is a great use case for this technique. That said, like all methods in this section, discourse analytics is time-consuming as the data needs to be analyzed until no new insights emerge.
- **Grounded theory analysis**: The grounded theory approach aims at creating or discovering a new theory by carefully testing and evaluating the data available. Unlike all other qualitative approaches on this list, grounded theory analysis helps in extracting conclusions and hypotheses from the data, instead of going into the analysis with a defined hypothesis. This method is very popular amongst researchers, analysts, and marketers as the results are completely data-backed, providing a factual explanation of any scenario. It is often used when researching a completely new topic or with little knowledge as this space to start from the ground up.

Quantitative Data Interpretation Method

The quantitative data interpretation method is used to analyze quantitative data, which is also known as numerical data. This data type contains numbers and is therefore analyzed with the use of numbers and not texts.

Quantitative data are of 2 main types, namely; discrete and continuous data. Continuous data is further divided into interval data and ratio data, with all the data types being numeric. There are few certainties when it comes to data analysis,it involves the use of statistical modeling such as standard deviation, mean and median. Some of the statistical methods used in analyzing quantitative data are highlighted below:

- **Mean:** a mean represents a numerical average for a set of responses. When dealing with a data set (or multiple data sets), a mean will represent a central value of a specific set of numbers. It is the sum of the values divided by the number of values within the data set. Other terms that can be used to describe the concept are arithmetic mean, average and mathematical expectation.
- **Standard deviation:** this is another statistical term commonly appearing in quantitative analysis. Standard deviation reveals the distribution of the

responses around the mean. It describes the degree of consistency within the responses; together with the mean, it provides insight into data sets.

- **Frequency distribution:** this is a measurement gauging the rate of a response appearance within a data set. When using a survey, for example, frequency distribution, it can determine the number of times a specific ordinal scale response appears (i.e., agree, strongly agree, disagree, etc.). Frequency distribution is extremely keen in determining the degree of consensus among data points.

Typically, quantitative data is measured by visually presenting correlation tests between two or more variables of significance. Different processes can be used together or separately, and comparisons can be made to ultimately arrive at a conclusion. Other signature interpretation processes of quantitative data include:

- **Regression analysis:** Essentially, it uses historical data to understand the relationship between a dependent variable and one or more independent variables. Knowing which variables are related and how they developed in the past allows you to anticipate possible outcomes and make better decisions going forward. For example, if you want to predict your sales for next month you can use regression to understand what factors will affect them such as products on sale, and the launch of a new campaign, among many others.
- **Cohort analysis:** This method identifies groups of users who share common characteristics during a particular time period. In a business scenario, cohort analysis is commonly used to understand customer behaviors. For example, a cohort could be all users who have signed up for a free trial on a given day. An analysis would be carried out to see how these users behave, what actions they carry out, and how their behavior differs from other user groups.
- **Predictive analysis:** As its name suggests, the predictive method aims to predict future developments by analyzing historical and current data. Powered by technologies such as artificial intelligence and machine learning, predictive analytics practices enable businesses to identify patterns or potential issues and plan informed strategies in advance.
- **Prescriptive analysis:** Also powered by predictions, the prescriptive method uses techniques such as graph analysis, complex event processing, and neural networks, among others, to try to unravel the effect that future decisions will have in order to adjust them before they are actually made. This helps businesses to develop responsive, practical business strategies.
- **Conjoint analysis:** Typically applied to survey analysis, the conjoint approach is used to analyze how individuals value different attributes of a product or service. This helps researchers and businesses to define pricing, product features, packaging, and many other attributes. A common use is menu-based conjoint analysis in which individuals are given a "menu" of options from which they can build their ideal concept or product. Through this analysts can understand which attributes they would pick above others and drive conclusions.

- **Cluster analysis:** Last but not least, cluster is a method used to group objects into categories. Since there is no target variable when using cluster analysis, it is a useful method to find hidden trends and patterns in the data. In a business context clustering is used for audience segmentation to create targeted experiences, and in market research, it is often used to identify age groups, geographical information, and earnings, among others.

Ladder of Inference

The ladder of inference is used to analyze the thinking process resulting in actions taken based on reality and facts. By using the ladder of inference, one can evaluate if they are jumping to conclusions or challenge other people's decisions to ensure they were made based on facts and data.

The ladder of inference is a decision-making model to assist the problem solver in reflecting on the various steps taken to make a decision based on facts and reasoning. The purpose of the ladder of inference is to understand the intentions and motivations behind actions and avoid jumping to conclusions.

The seven steps of the ladder of inference are (1) Observe data (2) selecting facts, (3) interpreting facts, (4) assumptions, (5) conclusions, (6) beliefs, and (7) actions. Each of the seven steps represents the cognitive processes involved from making observations to taking actions. This process can happen very quickly.

1. Observe data

The first step in the ladder of inference is to observe the reality and facts around you. This includes all the observable data, such as people's words, tone and body movements, a news report, results from a survey, etc.

2. Select data

The second step in the ladder of inference is to select the data you will process. Often you cannot pay attention to all the available facts and reality and tend to make conscious choices about what specific information to choose and what to ignore. This selection process mainly occurs subconsciously.

3. Interpret meanings

In the third rung of the ladder of inference, you tend to assign meaning to the information based on your own past experiences, beliefs and biases.

4. Make assumptions

Once you select and assign meaning to the data, you tend to make assumptions—this is step four, the midpoint of the ladder. At this stage of the ladder of inference, you might be quick to jump to conclusions about what's happening and ignore all the other facts and realities

5. Draw conclusions

Step five in the ladder of inference is when you begin to draw conclusions. Based on your interpretations and assumptions, you tend to explain why a particular event is happening and draw firm conclusions. As in the previous stage of the ladder (assumptions), all facts and realities are not considered to arrive at this conclusion.

6. Adopt beliefs

At the sixth step in the ladder of inference, you might adopt beliefs about situations that are associated with your conclusions. It is quite common to observe that these beliefs tend to shape your future judgments about similar scenarios.

7. Take action(s)

The seventh rung and top of the ladder of inference is when you take action. Based on your evaluation of the situation and the assumptions gathered, you decide what to do and tend to take action accordingly.

Your assumptions, beliefs and values influence how you select and interpret data. These interpretations and decisions will impact the decisions you make when you're in a similar situation the next time.

This action between the ladder of inference's second step (selected data) and the sixth step (beliefs) is called the reflexive loop, which is a cycle that leads you to constantly act on biased perceptions and limited information about a situation.

Methods of Communicating and Displaying Analysed Data

Having analysed the data that you collected through either quantitative or qualitative method(s), the next task is to present your findings to your readers. The main purpose of using data display techniques is to make the findings easy and clear to understand, and to provide extensive and comprehensive information in a succinct and effective way. There are many ways of presenting information. The choice of a particular method should be determined primarily by your impressions/knowledge of your likely readership's familiarity with the topic and with the research methodology and statistical procedures. If your readers are likely to be familiar with 'reading' data, you can use complicated methods of data display; if not, it is wise to keep to simple techniques. Although there are many ways of displaying data, this chapter is limited to the more commonly used ones. There are many computer programs that can help you with this task.

Broadly, there are four ways of communicating and displaying the analysed data. These are:

- text;
- tables;
- graphs; and
- statistical measures.

Because of the nature and purpose of investigation in qualitative research, text becomes the dominant and usually the sole mode of communication. In quantitative studies the text is very commonly combined with other forms of data display methods, the extent of which depends upon your familiarity with them, the purpose of the study and what you think would make it easier for your readership to understand the content and sustain their interest in it. Hence as a researcher it is entirely up to you to decide the best way of communicating your findings to your readers.

Text

Text, by far, is the most common method of communication in both quantitative and qualitative research studies and, perhaps, the only method in the latter. It is, therefore, essential that we know how to communicate effectively, keeping in view the level of understanding, interest in the topic and need for academic and scientific rigor of those for whom we are writing. Our style should be such that it strikes a balance between academic and scientific rigor and the level that attracts and sustains the interest of our readers.

Table

Other than text, tables are the most common method of presenting analyzed data. Tables offer a useful means of presenting large amounts of detailed information in a small space. They can dramatically clarify text, provide visual relief, and serve as quick point of reference. It is, therefore, essential for beginners to know about their structure and types.

A table has five parts

- Title – This normally indicates the table number and describes the type of data theυ table contains. It is important to give each table its own number as we will need to refer to the tables when interpreting and discussing the data. The tables should be numbered sequentially as they appear in the text. The description accompanying the table number must clearly specify the contents of that table.
- Stub – The subcategories of a variable, listed along the y-axis (the left-hand column of the table). The stub, usually the first column on the left, lists the items about which information is provided in the horizontal rows to the right. It is a vertical listing of categories or individuals about which information is given in the columns of the table.
- Column headings – The subcategories of a variable, listed along the x-axis (the top of the table).
- Body – The cells housing the analyzed data.
- Supplementary notes or footnotes – There are four types of footnote: source notes;υ other general notes; notes on specific parts of the table; and notes on the level of probability. These notes should be identified at the bottom of the table.

Graph

Graphic presentations constitute the third way of communicating analysed data. Graphic presentations can make analysed data easier to understand and effectively communicate what it is supposed to show. One of the choices you need to make is whether a set of information is best presented as a table, a graph or as text. The main objective of a graph is to present data in a way that is easy to understand and interpret, and interesting to look at. Your decision to use a graph should be based mainly on this consideration: 'A graph is based entirely on the tabled data and therefore can tell no story that cannot be learnt by inspecting a table. However, graphic representation often makes it easier to see the pertinent features of a set of data'.

Graphs can be constructed for every type of data — quantitative and qualitative — and for any type of variable (measured on a nominal, ordinal, interval or ratio scale). There are different types of graph, and your decision to use a particular type should be made on the basis of the measurement scale used in the measurement of a variable. It is equally important to keep in mind the measurement scale when it comes to interpretation. It is not uncommon to find people misinterpreting a graph and drawing wrong conclusions simply because they have overlooked the measurement scale used in the measurement of a variable. The type of graph you choose depends upon the type of data you are displaying. For categorical variables you can construct only bar charts, histograms or pie charts, whereas for continuous variables, in addition to the above, line or trend graphs can also be constructed. The number of variables shown in a graph are also important in determining the type of graph you can construct.

When constructing a graph of any type it is important to be acquainted with the following points:

- A graphic presentation is constructed in relation to two axes: horizontal and vertical. The horizontal axis is called the 'abscissa' or, more commonly, the x-axis, and the vertical axis is called the 'ordinate' or, more commonly, the y-axis.
- if a graph is designed to display only one variable, it is customary, but not essential, to represent the subcategories of the variable along the x-axis and the frequency or count of that subcategory along the y-axis. The point where the axes intersect is considered as the zero point for the y-axis.
- a graph, like a table, should have a title that describes its contents. the axes should be labelled also.
- A graph should be drawn to an appropriate scale. it is important to choose a scale that enables your graph to be neither too small nor too large, and your choice of scale for each axis should result in the spread of axes being roughly proportionate to one another.

Statistical measures

Statistical measures are extremely effective in communicating the findings in a precise and succinct manner. Their use in certain situations is desirable and in some it is essential, however, you can conduct a perfectly valid study without using any statistical measure.

There are many statistical measures ranging from very simple to extremely complicated. On one end of the spectrum you have simple descriptive measures such as mean, mode, median and, on the other; there are inferential statistical measures like analysis of variance, factorial analysis, multiple regressions.

Because of its vastness, statistics is considered a separate academic discipline and before you are able to use these measures, you need to learn about them. Use of statistical measures is dependent upon the type of data collected, your knowledge of statistics, the purpose of communicating the findings, and the knowledge base in statistics of your readership. Before using statistical measures, make sure the data lends itself to the application of statistical measures, you have sufficient knowledge about them, and your readership can understand them.

Reference

Kothari C.R. 2004. Research Methodology Methods and Techniques. Second edition: New Age International Publishers

Singh G.B. 2011. Research Methodology (Advanced Techniques with Statistical Methods). Paradise Publishers

Daivadeenam P.D, Somani L.L. 2004. Research Methodology in Extension Education. Agrotech Publishers

Multiple Choice Questions

1. **____________ is the process of reviewing data through some predefined processes.**

   a) task analysis b) **data interpretation**

   c) operational system d) none of the above

2. **A research method used to identify frequencies and recurring words, subjects and concepts in image, video, or audio content**

   a) content analysis b) data analysis

   c) system analysis d) thematic analysis

3. **__________ is used to analyze how individuals value different attributes of a product or service.**

   a) thematic analysis b) cohort analysis

   c) predictive analysis **d) conjoint analysis**

4. **It is used to understand how context can affect the way language is carried out and understood.**

    a) narrative analysis b) conjoint analysis

    c) discourse analysis d) grounded theory analysis

5. **Steps in ladder of inference**

    a) 8 b) 6

    c) 7 d) 10

6. **__________, by far, is the most common method of communication in both quantitative and qualitative research studies and, perhaps, the only method in the latter.**

    a) text b) pie chart

    c) graph d) excel

7. **__________ uses historical data to understand the relationship between a dependent variable and one or more independent variables**

    a) conjoint analysis **b) regression analysis**

    c) predictive analysis d) none of the above

One Mark

1. Define data interpretation
2. Methods of data interpretation
3. Define ladder of inference
4. Steps in ladder of inference
5. Grounded theory analysis
6. Methods of communicating and displaying analysed data
7. Brief note on table
8. Short note on cluster analysis
9. Quantitative data interpretation method
10. Qualitative data interpretation method

Chapter 30

Citation Practices and Referencing Styles

Citation

- ❖ A citation is a reference to any item (book, journal,articles, dissertation, archival manuscript, newspaper editorial, report, website, musical composition, etc.) which clearly identifies the sources in which the full-text on the item is to be found.
- ❖ It provide sufficient information to acknowledge the author and locate the item
- ❖ The information in a citation is fairly standard, but the order may change depending on the type of item being cited and the style manual you are using it includes,
  - ▷ Author
  - ▷ Article title
  - ▷ Publication information (i.e., Title, date, volume, issue, Pages etc.)
  - ▷ For online resource (the permalink/PURL/stable URL)

Citation Manager

- ❖ Citation managers (also known as citation management software, reference managers etc.), are available to help you collect, organize, cite and share references and sources (e.g. full-text articles).

Benefits of Citation Managers

- ❖ Manually formatting citations can be very time intensive! Familiarizing yourself with a citation manager early in your academic career can help keep you organized and can save you large amounts of time by obviating the need to manually format your citations.
- ❖ In addition to being a great tool for academic writing, citation managers can also help you collect, organize and store information sources for other purposes.

Citation managers allow you to:

- ❖ Manage all of your sources in one place.
- ❖ Upload and store full-text PDFs and other file types for your sources.

- Generate formatted bibliographies in desired format.
- Share collections with colleagues to collaborate in private or open groups.
- Facilitate the removal of duplicate citations
- Install plug ins for word processors that allow you to insert citations from the reference manager as you write
- Organize your references into meaningful collection folders that may serve a variety of purposes

How Citation Managers Work

- Collect: sources of information on your topic
- Import: citation information for your sources to a citation manager
- Upload: PDF's and other files to attach to your sources (optional)
- Organize: your sources into folders (optional)
- Cite-while you write: in desired citation style using Word plugins
- Generate bibliographies: in desired citation style

There are many different citation management programs available and many offer a freely available option.

The three most popular citation managers EndNote, Mendeley and Zotero, all offer an online and desktop version and provide a freely available option.

Common Referencing Styles in Social Research –MLA, APA, Chicago Manual Style

A citation is a way of giving credit to individuals or their creative and intellectual works that you utilized to support your research. It can also be used to locate particular sources and combat plagiarism. Typically, a citation can include the author's name, date, location of the publishing company, journal title, or DOI (Digital Object Identifier)

- A citation style is a set of rules on how to cite sources in academic writing.
- Citation style guidelines are often published in an official handbook containing explanations, examples, and instructions. The most common citation styles are the following:
  - APA (American Psychological Association) – used by education, Psychology and Sciences
  - MLA (Modern Language Association) – used by Humanities, Literature and Languages
  - Chicago manual style
  - Chicago author-date in the sciences.
- Different citation styles have different rules for in-text citations, reference list entries, and (sometimes) the formatting of your paper. The differences can be very subtle, so it's important to carefully check the rules of the style you are using.

Types of in-text citation

- When you refer to a source (for example, by quoting or paraphrasing), you have to add a brief citation in the text. There are three main types of citation:
  - ▷ Parenthetical citation: You put the source reference in parentheses directly in your text. This usually includes the author's last name along with the publication date and/or the page number.
  - ▷ Note citation: You put the source reference in a footnote or endnote.
  - ▷ Numeric citation:You number each of your sources in the reference list and use the correct number when you want to cite a source.

Citation style	*Disciplines*	*Type of citation*
MLA	Humanities	Parenthetical (author-page number)
APA	Psychology, education, social sciences	Parenthetical (author-date)
Chicago A	History, humanities	Notes
Chicago B	Sciences, social sciences, humanities	Parenthetical (author-date)

MLA STYLE

- MLA style citations are commonly used by students and academics in the humanities.
- To cite sources in MLA style, you need
  - ▷ In-text citations that give the author's last name and a page number.
  - ▷ A list of Works Cited that gives full details of every source.
- Make sure your paper also adheres to MLA format: one-inch margins, double spacing, and indented paragraphs, with an MLA style heading on the first page.

MLA Works Cited list

- The Works Cited list is where you give full details of all sources you have cited in the text. Other citation styles sometimes call this the "reference list" or "bibliography." An annotated bibliography is slightly different.
- Each Works Cited entry follows a template of nine core elements that all provide information about the source.
- Core elements of an MLA Works Cited entry

1 author	Johnson, David.
2 authors	Johnson, David, and Valerie Smith

Author. "Source Title." Container Title, Other contributors, Version, Number, Publisher, Publication date, Location.

The nine core elements of MLA citations

1. Author

- Begin each source entry with the name of the author(s) or creator(s). The name of the first author is always inverted (Last name, First name).
- When a source has two authors, the second author's name is shown in the normal order (First name Last name).
- For sources with three or more authors, state only the first author's name, followed by "et al."
- The author of a source is not necessarily a person; it can also be an organization. If so, simply use the name of the organization.

2. Title

- Always include the full title of the source, including subtitles (separated by a colon and space).
- Use title case capitalize all words apart from conjunctions, prepositions, and articles. If there is no title, give a short description of the source, with normal sentence case capitalization.
- The styling of the title depends on the type of source:
  - Italics when the source is self-contained (e.g. a whole book, movie or website).
  - Quotation marks when the source is part of a larger whole (e.g. a chapter of a book, a page on a website, or an article in a journal).
  - No styling when describing a source without a title.

3. Container

- A container is the larger work that the source you're citing appears in. For example, a chapter is part of a book, a page is part of a website, and an article is part of a journal.
- If the source you're citing is a self-contained whole (e.g. a whole book), leave out this element

4. Other contributors

- Contributors are added right after the container title and always end with a comma. Use a description like "translated by," "directed by," or "illustrated by" to indicate the role of the contributor. For example:
- Latour, Bruno. *Politics of Nature: How to Bring the Sciences into Democracy*. Translated by Catherine Porter, Harvard UP, 2004.

5. Version

- When there is more than one version of a source, you should include the version you used. For example, a second-edition book, an expanded version of a collection, or a director's cut of a movie would require the version to be included:

- ❖ Porter, Michael E. Competitive Strategy: Techniques for Analyzing Industries and Competitors. 2nd ed., Simon and Schuster, 1998.
- ❖ Columbus, Chris, director. Harry Potter and the Chamber of Secrets. director's cut, Warner Bros., 2002.

6. Number

- ❖ Sources such as journal articles ("vol. 18"), magazines ("no. 25") and TV shows ("season 3, episode 5") are often numbered. If your source has numbered parts, include this in the source entry:
- ❖ Wieseky, Jan, et al. "Willing to Pay More, Eager to Pay Less: The Role of Customer Loyalty in Price Negotiations." Journal of Marketing, vol. 68, no. 6, 2014, pp. 17–37.
- ❖ It is also possible for a source to have an edition, volume, and number. Just separate them using commas.

7. Publisher

- ❖ Book and movie citations always include the publisher element. The publisher is the company responsible for producing and distributing the source—usually a book publisher (e.g. Macmillan or Oxford UP) or a movie production company (e.g. Paramount Pictures or Warner Bros).
- ❖ Note that "University press" is abbreviated to "UP" in a works cited entry. Example: The University of Minnesota press becomes "U of Minnesota P", Oxford University press becomes "Oxford UP"

8. Publication date

- ❖ When available, always include the publication year. If you also know the month, day, or even time of publication, you can include this if it helps the reader to locate the source. Date ranges are also possible. For example:
- ❖ 2003
- ❖ 25 Jan. 2019
- ❖ 14 Aug. 2017, 4:45 p.m.
- ❖ Jan. 2017–Apr. 2018

9. Location

What you include in the location element depends on the type of source you are citing:

- ❖ Book chapter: Page range of the chapter (e.g. pp. 164–180.)
- ❖ Web page: URL, without "https://" (e.g. www.scribbr.com/mla-style/quick-guide/.)

- Journal article: DOI, with "https://"—or stable URL, without: (e.g. https://doi.org/10.1080/02626667.2018.1560449. or www.jstor.org/stable/43832354.)
- Physical object or live event: Name of the location and city (e.g. Moscone Center, San Francisco. or The Museum of Modern Art, New York.)

The standard MLA book citation format is as follows:

Author's last name, first name. Title. Title of container, Contributors, Version, Numbers, Publisher, Year of Publication

- Example: Fitzgerald, F. Scott. The Great Gatsby. Scribner, 2004
- Journal example: Grin, Lea. "Science Fiction Works for the Development of the Aerospace Sector." The Popularisation of Space, vol. 41, Aug. 2017, pp. 42-47.
- Magazine/Newspaper example: Smith, John. "Journal
- Obama inaugurated as President." Time, 21 Jan. 2009: 21-23. Print.

APA Style

- APA Style was created by the American Psychological Association and was originally used in psychology and the social sciences. Today many other disciplines also use it. APA's citation style uses an author-date system of parenthetical citation.
- APA Style citation example (book)

APA in-text citation

- The book uses success stories from the authors' own lived experiences to make the case for a creative revolution in the business world (Stewart & Simmons, 2010, p. 22

APA reference list entry

- Stewart, D., & Simmons, M. (2010). The business playground: Where creativity and commerce collide. New Riders Press.
- Note: This example follows the 7th edition of the APA Publication Manual, released in October 2019.

Different Scenarios for Writing References According to APA Format

Websites - Authors/ Organizations/ without author/ without date

1. With Author

- Author surname, Initial. (Year, Month date). Title in italics. Website. URL.
- Example: Sparks, D. (2015, September 15). Practicing mindful experience. Mayo clinic.https://newsnetwork.mayoclinic.org/discussion/mayo-mindfulness-practicing-mindfulness

2. With 2 Authors

- Author surname, Initial., & Author Surname, Initial.(other same as above)
- Klimonske, R., & Palmer, S. (2015, September 15). Practicing mindful experience. Mayo clinic. https://newsnetwork.mayoclinic.org/discussion/mayo-mindfulness-practicing-mindfulness

3. No Author but any Group 'S or Organization's Work

- National Institutes of Mental Health. (2015, May). Anxiety disorders
- http://www.nimh.nih.gov/health/topics/anxiety-disorders/index.s

Journals - General/ 2 to 20 authors/ more than 21 authors..

1. General format

- Author last name, Initial. (Year of publishing). Article title. Journal title in Italics, Volume (issue), Page number. URL
- Example: Scroggins, W. A., Thomas, S .L., & Morris, J. A. (2008). Psychological testing in Personnel selection, Part II: The refinement of methods and standards in employees selection. Public Personnel Management, 37(2),185-199.
- https://doi.org/10.1037/1061-4087.45.2.10

2. Two or more authors (up to 20 authors in journals)

- Example: Klimonske, R., & Palmer, S. (1993). The ADA and the hiring process in organizations. Consulting psychology journal: Practice and Research, 45(2), 10-36. https://doi.org/10.1037/1061-4087.45.2.10

3. More than 21 authors

- Format: If an article has 21 authors or more, list the first 19 authors, then insert an ellipsis (...) and then the last name and first initials of the last author.
- Example: Wolchik, S. A., West, S. G., Sandler, I. N., Tein, J., Coatsworth, D., Lengua, L., Johnson, A., Ito, H., Ramirez, J., Jones, H., Anderson, P., Winkle, S., Short, A., Bergen, W., Wentworth, J., Ramos, P., Woo, L., Martin, B., Josephs, M., ... Brown, Z. (2005). Study of the Brain. Psychology Journal, 32(1), 1-15. https://doi.org/10.1037/1061-4087.45.2.10

News/Magazine Article:

- Author(s). Note: List each author's last name and initials as Author, A. A., Author, B. B., & Author, C. C. Use an ampersand (&) before the final author's name.
- (Year, Month Date). Note: You do not need to abbreviate the month.
- Title of the article. Note: For works that are part of a greater whole (e.g. articles, chapter), use sentence case. Only the first letter of the title and subtitle and proper nouns are capitalized.

- Title of the Newspaper or Publication. Note: Italicize and capitalize each word in the publication.

4. URL

- Kennedy, M. (2018, October 15). To prevent wildfires, PG&E pre-emptively cuts power to thousands in California. NPR. https://www.npr.org/2018/10/15/657468903/to-prevent-wildfires-pg-e-preemptively-cuts-power-to-thousands-in-california

Book chapters with editors:-

- Author(s). Note: List each chapter author's last name and initials as Author, A. A., Author, B. B., & Author, C. C. Use an ampersand (&) before the final author's name.(Year).
- Title of the chapter. Note: For works that are part of a greater whole (e.g. articles, chapter), use sentence case. Only the first word of the title and subtitle and proper nouns are capitalized.
- In Editor(s), Note: List each editor's last name and initials as A. A. Editor, B. B. Editor, & C. C. Editors, include (Ed.) or (Eds.) in parentheses, and end with a comma.
- Title of the book Note: For works that stand alone (e.g. books, reports), italicize the title. Only capitalize the first word of the title and subtitles and any proper nouns.
- Publisher. Note: You do not need to include the publisher location or databases where you retrieved it.
- McCormack, B., McCance, T., & Maben, J. (2013). Outcome evaluation in the development of person-centred practice. In B. McCormack, K. Manley, & A. Titchen (Eds.), Practice development in nursing and healthcare (pp. 190-211). John Wiley & Sons

Chicago style

- Chicago style is published by The Chicago Manual of Style. There are two variations:
- Chicago A: You cite the source in a footnote or endnote.
- Chicago B: You use a parenthetical author-date citation in the text.
- Chicago A example (book)

Chicago in-text citation

- The book uses success stories from the authors' own lived experiences to make the case for a creative revolution in the business world.

Chicago style footnote

- [1] Dave Stewart and Mark Simmons, The Business Playground: Where Creativity and Commerce Collide (Berkeley: New Riders Press, 2010).

Chicago bibliography entry

- Stewart, Dave and Mark Simmons. The Business Playground: Where Creativity and Commerce Collide. Berkeley: New Riders Press, 2010.

Chicago B example (book)

Chicago in-text citation

- The book uses success stories from the authors' own lived experiences to make the case for a creative revolution in the business world (Stewart and Simmons 2010, 22).

Chicago reference list entry

- Stewart, Dave and Mark Simmons. 2010. The Business Playground: Where Creativity and Commerce Collide. Berkeley: New Riders Press.

Chicago Style GuideChicago citation examples

Chicago Notes and Bibliography`	
Footnote or endnote format	Author first name last name, Title of Book (Place of publication: publisher, year), page number(s).
Full note (first mention)	Albert Einstein, The Meaning of Relativity (Princeton: Princeton University Press, 1923), 44–45.
Short note (subsequent mentions)	Einstein, The Meaning of Relativity, 89.
Bibliography format	Author last name, first name. Title of Book. Place of publication: publisher, year.
Bibliography example	Einstein, Albert. The Meaning of Relativity. Princeton: Princeton University Press, 1923.

Chicago Author-Date	
In-text citation format	(Author last name year, page number(s))
In-text citation example	(Einstein 1923, 44–45)
Reference list format	Author last name, first name. Year. Title of Book. Place of publication: publisher.
Reference list example	Einstein, Albert. 1923. The Meaning of Relativity. Princeton: Princeton University Press.

Chicago journal article citation	
Footnote or endnote format	Author first name last name, "Title of Article," Name of Journal volume, no. issue (month and year): page number(s). DOI if applicable.

Full note (first mention)	Morris Dickstein, "A Literature of One's Own: The Question of Jewish Book Awards." Princeton University Library Chronicle 63, no. 1–2 (Winter 2002): 71. https://doi.org/10.25290/prinunivlibrchro.63.1-2.0070.
Short note (subsequent mentions)	Dickstein, "A Literature of One's Own," 73.
Bibliography format	Author last name, first name. "Title of Article." Name of Journal volume, no. issue (month/season year): page range of article. DOI if applicable.
Bibliography example	Dickstein, Morris. "A Literature of One's Own: The Question of Jewish Book Awards." Princeton University Library Chronicle 63, no. 1–2 (Winter 2002): 70–74. https://doi.org/10.25290/prinunivlibrchro.63.1-2.0070.

Chicago website citation (no author or date)	
Footnote or endnote format	"Title of Page," Website, accessed month date, year, URL.
Full note (first mention)	"About the UvA," University of Amsterdam, accessed July 24, 2018, http://www.uva.nl/en/about-the-uva.
Short note (subsequent mentions)	"About the UvA."
Bibliography format	Website. "Title of Page." Accessed month date, year. URL.
Bibliography example	University of Amsterdam. "About the UvA." Accessed July 24, 2018. http://www.uva.nl/en/about-the-uva.

Chicago newspaper citation	
Footnote or endnote format	Author first name last name, "Title of Article," Name of Publication, month date, year, page number or URL.
Full note (first mention)	Alex Marshall, "Graphic Novel in Running for Man Booker Prize for First Time," New York Times, July 23, 2018, https://www.nytimes.com/2018/07/23/books/booker-prize-graphic-novel-ondaatje.html.
Short note (subsequent mentions)	Marshall, "Graphic Novel in Running for Man Booker Prize."
Bibliography format	Author last name, first name. "Title of Article." Name of Publication, month date, year. URL if applicable.
Bibliography example	Marshall, Alex. "Graphic Novel in Running for Man Booker Prize for First Time." New York Times, July 23, 2018. https://www.nytimes.com/2018/07/23/books/booker-prize-graphic-novel-ondaatje.html.

References

Citation Styles Guide | Which Citation Style Should You Use? (scribbr.com)

Citation Managers - Citing and Citation Managers - Research Guides at Queen's University Library (queensu.ca)

MLA Citation Generator (Free) | Verified by Experts | Scribbr

APA Citation Generator (Free) | Verified by Experts | Scribbr

Chicago Style Citation Guide | Templates & Citation Examples (scribbr.com)

Short Questions

1. Define citation
2. What are citation managers?
3. List down the common citation styles
4. Briefly describe the types of citation
5. Explain briefly on MLA style
6. Explain the core elements of MLA style
7. Define APA style
8. Cite an example for APA style
9. Differentiate between the Chicago A and B style
10. Cite an example for Chicago style

Multiple Choice Questions

1. _______ is a reference to any item which clearly identifies the sources in which the full-text on the item is to be found

a) **Citation**

b) Journal

c) Thesis

d) None of the above

2. Citation managers are also known as _______

a) Citation management software

b) Reference managers

c) **Both a and b**

d) None of the above

3. _______ help you to collect, organize, cite and share references and sources

a) Citation

b) **Citation manager**

c) Journal

d) Articles

3. **A _____ is a set of rules on how to cite sources in academic writing**
   a) Citation
   b) Citation manager
   c) Reference manager
   d) **Citation style**

4. **MLA style citations are commonly used by students and academics in the_____**
   a) **Humanities**
   b) Psychology
   c) Medicine
   d) All of the above

5. **_____ citation style uses an author-date system of parenthetical citation**
   a) MLA
   b) **APA**
   c) Chicago A
   d) Both a and b

6. **Chicago style is published by _____**
   a) American Psychological Association
   b) Modern Language Association
   c) **The Chicago Manual of Style**
   d) None of the above

7. **For sources with three or more authors, the first author's name is followed by**
   a) **et al.**
   b) Last name
   c) First name
   d) Version

8. **_____ allow you to generate formatted bibliographies in desired format**
   a) Citation styles
   b) **Citation manager**
   c) Citation
   d) None of the above

9. **_____ citation style uses author-page system of parenthetical citation**
   a) Chicago A
   b) Chicago B
   c) APA
   d) **MLA**

Index

Q

R

S

T

U

V

W

X

Y